BEYOND THE BREAK

BEYOND THE BREAK

The Return
Book I

JEREMY MICHAEL BECK

Jeremy Michael Beck Publishing

First printing: 2022.

ISBN: 979-8-218-04517-3

Jeremy Michael Beck Publishing
https://jeremybeckoffcial.wordpress.com

CHAPTERS

CHAPTER I

Oval Office, 2029

While alone and deep in thought, I had been gazing into the eyes staring back at me in the windowpane of the Oval Office. They looked ancient, the brow wizened by the mind that remembered all they had seen – yet they remained sharp and a vivid sky blue.

"Mr. President?" came a hesitant voice beyond me.

I was suddenly aware of the many bodies in the room. In the other panes, I saw the dull eyes of those amassing behind me. They were at the distance of the far wall. The light coming in from the hallway revealed matted locks of hair and pitted faces.

Calling this an all-nighter would not give due justice to the stress felt by these staffers, looking for guidance in what was becoming an impossible situation. It was late, but the work of the White House never stopped.

The events of the past few hours had come as a great surprise to everyone, yet they were expectant of me. They wondered *who* they could trust – they wondered if we could trust each *other*. Beliefs they knew as truths were now coming into question.

They wondered if this meant war.

As one day had become another, in the deepest of darkness, a coup was enacted to overthrow the government in North Korea. The military had become such a tool for the individual they would no longer comply.

His home had been raided, and they dragged him into the streets. As the crowd retreated from the beaten and broken man, his blood witnessed to shimmer in the sunlight – though otherwise translucent.

The people believed him to be an alien life form that had infiltrated the highest levels of the North Korea government. It was unknown if they had human allies – or otherwise – or if they had compromised other world powers.

As I turned to face those gathered in the Oval Office, I believed there was no one in the room that could have prepared for this.

"The decision to respond has to be made by Congress," I replied. "What's important is that I address the *people* – assure them of their safety, the safety of our borders, and our ability to defend ourselves, should the need arise. Have Harry prepare some statements." I looked around the room at the gathering of concerned souls. "That is all." I turned to gaze out the window.

The past two years, they had learned that my words meant much, though I would speak little.

They filed out of the room subconsciously expressing their new found paranoia by exchanging quick and piercing glances. They would inform those needed to be present for the morning's proceedings, and we would adjourn to the Situation Room. Believing I was alone to prepare myself, I turned back to the window, looking out across the White House lawn and beyond.

There were dark clouds on the horizon – lightning, but no rain.

The Vice President, members of my Cabinet, Security Advisors, the Secretary of Defense, and the Joint Chiefs of Staff, would share the halls with many senators and other representatives. Dodging them would be other members of my Executive Office coordinating with reporters. They would move correspondence throughout the government alongside the Department of State communicating internationally.

Somewhere along that line, rumors would give rise to questions we must now answer.

A press briefing would follow these meetings, while my most trusted advisers and I would have a clandestine discussion about the true nature of this situation. We must decide what truths to expose to the American people and what to share with our allies.

We would focus the eyes of the nation in great scrutiny on our next step forward.

I heard a shuffling and noticed a presence that had remained in the room.

"Well, well, well... this should be interesting, eh?" came a voice from a shadowed form now sitting relaxed on the couch in the middle of the room, one arm draped over the back. "This really, what is it they say around here, throws a wrench in the gears? I hope you're not having second thoughts again, *Oscar*."

It was my Chief of Staff, one Jacqueline Stone.

A recent addition to my team and not one I would have chosen. She had just returned from an extended "vacation," and stood relaxed, her features betraying the serious nature of the situation. A stout woman and when she pushed aside her short dark hair, I glimpsed the scar behind her right ear, rarely visible and almost forgotten. She presented herself as the professional, and I had only become aware of the scar as she seemed to be more lax in her demeanor each day.

This infiltration was no news to the two of us, and perhaps our enterprise was nearing its end.

While the presumptive leader of this country, there were many whose voice carried much more persuasive power than my own. Some held their power with wealth, like those who had recruited Jaqueline. Their power came from position and ancient ranks much higher than my own in their grand scheme. They sent Jaqueline to ensure there was no change made to their plan – no delay to their timetable.

"None whatsoever, *Jackie*," I retorted. That put a wrinkle on her nose. She hated her chosen name – even more-so the intimate connotation of this shorter form. "I see that I won't be allowed any sleep tonight."

"Afraid not. There are important decisions to make, and they need to be made now. It was nice of you to let them think they could prepare your speech, though we both know you never read them," she remarked. "You will receive instruction tomorrow, but we will need your insight on how the population will react," she said smugly.

"You might wish to stop calling them 'the population.' It is not very convincing," I replied.

"Look, I'm new here, and none of us will be remaining much longer anyway, *Mr. President*. You have your title, you have your little plans. But do not forget who pulls the strings in this operation. You have superiors who see the big picture, *soldier*," she said in a tone of condescension. "You are no longer in control of this mission."

She then turned, pausing as if to hold the tension just a moment longer. She felt powerful in this room, alone with me, where she could openly flaunt her higher position. I was not the true politician, but she had watched as the others listened intently to my every word, awaited my praise, and trusted my intuition.

She watched, knowing she was the medium through which I received that intuition.

She did not turn to face me, but I could feel a smirk creep up the edge of her lips. She then exited the Oval Office and left me to my thoughts.

Truly, this was the moment I had waited over half a century for – with many compromises made along the way. I had placed myself at the forefront of these designs.

It was a responsibility I had desired. I knew there would be many obstacles – I had overcome them all. It was not always easy to follow this path. I had gained debts that still held my obligation and secrets kept ensuring my place here, where our sins began.

I sat at my desk to prepare for the days ahead. I pushed aside a document that had always piqued my interest. It was a recounting of what some believe to be the first UFO encounter on American soil in Aurora, Texas, 1897.

The remains of a craft were buried away in a well, and we know that the farmer that made the discovery developed many health problems related to aluminum poisoning and radiation exposure. They found a body in the wreckage which is said to be buried alongside an extraordinary object in the local cemetery.

I wondered how William McKinley would have responded to this claim, foregoing his assassination.

What followed were decades of rises and falls in the economy perpetrated by a financial reliance on war and internal chaos. Americans rose out of a depression through the industry of death machines – paid for by enemies and allies alike. The greatest wars and atrocities this world may ever produce define the era.

An era that was characteristically human.

Because of my station, I knew that first contact was not made until several years after the first genuine discovery of extraterrestrial life – the Roswell crash of 1947. However, I believed this Aurora report – whether true *or* false – created a predisposition to fear the possibility of extraterrestrial life.

It would not be long before I would begin receiving communications. It was important I be of sharp mind to collect any information I would need to construct my narrative. While I would be given major talking points, the delivery of the speech would be of my authorship. While grammar is important to appear the competent leader, what was truly moving was the inflection of your words and using powerful phraseology.

One had to infect an audience with a grand passion.

While much of the data I receive will be statistics, I would listen for personal stories from the future front. What I needed was a story the people could relate to. These are reports most would consider white noise.

I thought of my predecessors.

"Four score and seven years ago...," I began aloud.

A standard introductory statement that did no more than establish a time frame. Few words remembered for the man who spoke them and by that historical location known to every American man, woman, and child. It was the beginning of a relatively brief speech for such a terrible issue – slavery, and the ending of a bloody civil war – but spoken powerfully enough to reverberate throughout the ages.

"The only thing we have to fear is fear itself!" I continued in my best Delano.

Words meant to inspire the downtrodden citizens suffering most from the Great Depression. An inaugural introduction from the fresh face of a nation that wanted to lead America into a prosperous future. Restated were the promises of the campaign trail. This address to the people invoked the power of religious language, a great motivator in this country.

My time had finally arrived to take charge, and I thought of a speech written by the previous President.

The first woman to be elected to lead the United States, though loyal to the aliens, she referenced another powerful moment in American history. She gave her inaugural address in front of the Lincoln Memorial, where Marian Anderson was forced to sing after the Daughters of the American Revolution refused to allow a black artist to perform at Constitution Hall.

"And so it is fitting that we should gather in tribute to the will of the people, with 2020 vision! They saw the future is *clear*!" A very common expression, but when something is so simple and rings so true to the heart, the people need not hear more. These words would have rallied support for any campaign and carried her to a second term.

I only wished I could borrow them now.

Each of these speeches has a common theme that drew them to my mind – each was written by the speaker. The common person believes the cries that erupt from the mouths of their leaders are their own, but this is rarely the case. This is the practice at every level of authority, but my pride would not allow it. I hoped to glean some insight from my true comrades.

A signal was broadcast directly to a receiver under my desk, known only to myself and Jaqueline and transmitted through a secure dark room to the side of the Oval Office. I initiated a conference with my equals in this mission, now positioned as several other world leaders. We operated autonomously, with the occasional intervention from advisers who addressed *our* leadership – the Imprir, heads of the Militia Majora, the alien chain-of-command.

I was not the only one disturbed by our operative's deposition in North Korea.

"[If they had left us to our devices in the first place, we would not have been risking exposure!]" I heard in Hindi.

"[How could we have been compromised by such infiltration?]" came a voice in French. ["Are *they* not more thorough, better –"]

"Yes, yes, but it seems we are not so different after all," I said. "We are beyond wondering how this happened. We must now find a path to the future."

Usually I was the silent one, but no voice spoke in any language. I could hear only static as the many signals interfered with each other in the darkroom. The silence was finally broken by a cough from Jaqueline.

"We will have our orders. I have been told to terminate this conference." The lights went out on the switch board below my desk, interrupting a few confused queries. Jacqueline swayed over to the other side in her practiced gait.

"It would seem we are in a situation similar to that of our enemies. We don't know who to trust or who has been exposed. We don't know if anyone is helping these dissidents or revealed our intent." I sensed she was reluctant to tell me more.

"Is there any news you would like to share with me?" I asked knowingly.

She stared back at me and I could see there was no levity in her eyes. "China."

China was our most secure operation. While the United States had been the aliens early focus, this resulted from a chance crash landing on American soil. The *actual* power of these twelve nations was in China – greatest population with the most restricted access to the outside world. This made it much easier to limit influence from more open government culture.

This would cause change – change made our work very difficult.

The key to keeping a technologically advanced population sedated was controlling that technology and the flow of information through it. Most of our private data servers were hidden in China for this reason. The government had provided a system for monitoring all online traffic and used it to restrict information from getting in.

We wished to prevent information from getting out.

"Until we know if anyone else has been compromised, there will be no communications between China and the other teams. If any of them attempt to contact you, you are not to respond, understood?" she inquired, to which I gave a nod. "Your correspondence will be arbitrated through me, so continue writing your speech. After the show tomorrow, we shall head to SB-3 for an emergency meeting."

"And who are we sending to China?" I asked.

"Oh, don't worry about that now, *Mr. President.* You can't be expected to do everything..."

Again alone, I wondered what would become of my comrades. The exposure in North Korea was public and brutal, perpetrated by the military. Our plans in China had been foiled by the civilian population. They were not satisfied with their lack of freedoms even before we had restricted them further. A database had been hacked, leaking our strategy of replacing these world leaders. Before all records of it could be wiped from the internet, it made its way to the personal device of what seemed to be a consort of our imposed leadership.

Not all within the government were under our influence, and this information reached them as well. A scheme had been orchestrated to introduce an agent to the President of the People's Republic of China and blackmail him to get even closer. Our operative had kept her a secret even from us, truly a victim of the world's many vices. Had we known earlier, the agent would have been investigated and her plot surely discovered.

What would possess someone to risk such a sacrifice was beyond my comprehension.

"The best-laid plans of mice and men often go awry...," I began muttering to myself. I had memorized many quotes – rather than garnering their meaning – and it seemed appropriate. As I wandered across the Rose Garden, I thought of the days ahead. There were many faces I had not seen in many years I will see tomorrow – most of them with more than one name. I entered the White House proper and took to navigating the many corridors.

On my stroll, I was never free of the watchful eyes of the Secret Service. These were men and women I could trust, trained to see without looking, hear without bending an ear, and reveal nothing they observed. They had never disclosed our many secret meetings where we established our plans for mankind.

But they *are not men*, I thought.

I entered my personal study and picked up my pen. I prepared myself for a long night of writing, rewriting, and self-loathing. Speech was an art, and tonight I was the starving artist – the full moon, my muse. Usually, words came easily, flowing out of my brain through my arm and the pen onto the page in a deep black ink. I had to create a bold lie while revealing that which could no longer be denied.

Pandora's box has been flung open and would not easily be sealed.

I sat there for hours. Three times I had written an introduction, the ink light on the page from the lack of pressure and conviction. Three times I threw away a crumpled ball of paper. I returned the pen and paper to the right-side drawer of my desk and retired to bed.

Perhaps I would get some sleep tonight after all.

* * *

My dreams rarely strayed from photographic memories, but this night I found myself enveloped in a suffocating void. I felt pressure from every direction from unseen entities, bustling my body about in their sick dance. Crying out to any who would aid me, I shut my eyes tight and found an answer from within.

I opened my lucid eyes to find the void being filled with light battling the darkness away. I saw myself – dark hair over burgundy skin, glowing a bright maroon in light emanating from within my form. I began chanting in a language I did not recognize but understood:

One cannot change the past without the future – the present shall cease in all dimensions.

CHAPTER II

The Speech

Video of the public executions in North Korea had surfaced on the internet in morning newsfeeds. The plans discovered in China had also been disseminated to the public. The entire world was aware – and afraid – unable to trust family or friend.

They knew we were not alone.

These aliens were hostile. They had infiltrated their communities, and the world governments had been working with them for decades. The people turned on each other – every country was experiencing riots, and street justice was perpetrated in every community. Until one was bleeding, one could not tell alien apart from human.

"Mr. President, the Secret Service – D.C. *police* – cannot provide enough cover! We need to bring in the National Guard," asserted General Tom Wilkerson, my homeland security advisor.

The White House was under heavy lock-down due to the citizens amassing outside. They wanted proof that their leaders were not aliens to *more* than their country – that we were not compromised. In the Situation Room, things were no less hectic and none were shy to provide suggestions.

"Mr. President, we need a plan of action for our international interests. Things are getting out of hand and we will not know who is in control of what army and their nukes. I vote for a plan of *attack*," said another adviser.

"Why not provide the sample? It's no different than showing a birth certificate," added an analyst.

Jacqueline entered the room, and I knew it was time to end this discussion.

"We don't ask the President of the United States to provide evidence of their legitimacy, whether it be of his citizenship, or his species!" cried my Vice President, Richelle Dolan. "If we test him, what then? Do we test everyone in the government? Do we install checkpoints where any citizen must be pricked and prodded for their blood?" she continued. "Is that going to be the purpose of bringing in the National Guard, Tom?" she asked, turning to the General.

"The press briefing, sir," Jacqueline informed us. "First, we –,"

"If we must, Madame *Vice* President, to protect the security of our citizens –" Tom interjected, ignoring Jaqueline's presence. She did not appreciate this interruption or the condescending tone – in fact, she was his superior.

"I was *speaking*. And I believe Vice President Dolan will suffice."

"Y– Yes, of course," Tom apologized, stuttering.

"Police and soldiers in the streets will not make anyone feel more secure," I assured Tom. "We can*not* allow this to escalate. They need to feel like they can go about their daily lives without worry of an imminent invasion. These aliens will not interrupt the democratic process! We will proceed with what security we have. We will take no aggressive action while boasting our strength. We have to assure them of the capability of our defenses, that we have not been compromised, and that we *will* defend ourselves." I nodded and stood to leave the room. The others stood, and some gave a salute.

I exited the room, with my Chief of Staff one step behind me. We walked down the hallways of the West Wing and through the door to the Oval Office. We approached a bare spot on the wall with no picture hanging from it or any furnishings against it.

I kept an ID chip resembling a common SD card hidden in a shirt cuff. Retrieving it and placing it in a cartridge slot, I placed my thumb on a module that checked my DNA. The force field deactivated while retaining the illusion of a solid wall. I stepped through this false barrier without hesitation, but Jacqueline remained outside the hidden passage. She seemed to have urgent news, but after a few moments, I assumed I would remain unsupervised.

I approached the panel in the center of the room. A screen appeared in the space just in front of the wall and it was divided into twelve spaces. Several of them were only showing static – China, North Korea, and now Brazil, Germany. Some were simply blacked out. Only two members of my team were present in their own bunkers – France and Russia – but I heard only the English language here.

"How are things on your end?" I asked.

I could not see their faces, though I knew them well. We had all been raised together for these very roles. While every group has its rivalries, we were best defined as friends or comrades – closer than family.

"We have a hold on things here, Double-O," came the voice of Sergei Zuyev, my personal rival turned the closest ally among our unit. I did not mind the affectionate connotations of nicknames as much as Jacqueline. "Russia will live," he concluded.

"Good to hear. And you, Mathilde?" I inquired of the darkened visage from France, Mathilde Cordier. She had no handler – this was an alien. After an accident took the life of her charge – a loss still fresh to those of us who had served with the true Mathilde Cordier – she took on the role, forced to again change her form.

I hear the pain of this transformation is excruciating.

"We are blockaded and under siege throughout most of Paris. We still control the military and have confirmed that some of the instigators have come from Germany. It appears they have succeeded in Italy through collaboration of their armed forces to take control of the government houses," she said, confirming the reports that had come in throughout the night.

"Italy is not one of ours."

"It is not. Our intel suggests they will come for *you* next, Sergei." It seemed more and more likely few of our comrades would join us.

A new screen was activated and was receiving from Nigeria. This was where we held much of our ordinance, and it was reassuring to hear from them.

"I have been briefed on the situation and had my soldiers take the streets to prevent any sort of uprising," came the commanding voice of Adebowale Ikande. Nigerian leadership had offended the Imprir by denying resources the Militia Majora needed for a new project, so Adebowale was ordered to lead a military coup and had been in control for several years. He was also an alien.

Publicly, Nigeria and the United States were now enemies, though the two of us had been comrades even before our recruitment by the Majestic Twelve, the only two soldiers in this crew. The others were career politicians, the Imprir believing they would be more successful in filling these roles.

It seemed their numbers were dwindling while we remained in control.

"The missiles?" I was quick to ask.

"Our missiles are safe, but we hear we are missing a few codes," came a second voice, Adebowale's handler, whose name was unknown to me. "I shouldn't need to remind you we must have seven of the twelve launch codes to enact their plan."

"You do not. And I need not remind any of you we need to take control of this situation – now. The people are asking for proof that their leaders are not alien. I worry if we do not, they will turn on us and continue to turn on each other. We have hundreds of reports of random attacks, many of them fatal."

"We have the same reports," added Mathilde. "None of them resulted in the death of an *operative* – only humans. It seems many simply need an excuse to practice their violence. Sad, really, but it shall keep them busy. As for *your* identities, it is simple. We will provide the samples in a public show of honesty." She cut off her transmission, disclosing no more details.

I wished my remaining comrades well and exited the dark room. My materialization on the other side of the wall caused Jaqueline to jump.

"I'm sure there was more we could have discussed, or are our overlords becoming even more paranoid?"

Jaqueline did not turn to me as she grunted a fake laugh. "You have two hours. I hope you have some pretty words for the people. I wouldn't want you to appear *superfluous*."

"Should I tell them we're going to war?"

"No, we are not ready for phase two quite yet. We need to have stability – or at least the appearance of it – until we retrieve the codes." She looked up at me with a condescending smirk. "That is all for now, *Mr. President*." She then exited to the hallway.

I sat at my desk and thought of the friends I had lost in this endeavor. I took the pen in my hand and with nothing to show from the night before, I began anew. This time, I needed only one attempt.

* * *

As I approached the back of the stage, I could barely see through the wall of agents and police officers. On the other side of the barrier, however, I could *hear* the size of the crowd.

They did not sound satisfied.

I paced nervously and swallowed a pill. I had prepared my speech and would perform the test afterwards, to the dismay of Jacqueline. The Vice President was speaking presently. She had already taken her test in front of the gathered crowd, but this did not provide them any solace. The previously exposed conspirators were at the top of their respective hierarchies.

They were here for me.

"Mr. President, we're ready for you, sir," came the voice of a woman wearing a headset. She turned away as she covered the earpiece with her hand to better hear her next command. I headed towards the side of the stage and as I passed Richelle, she looked at me with a sweaty brow and shook her head.

"I warmed them up for ya," she joked, to hide her concern. We shook hands, and I continued up the steps and towards the podium with many microphones – like the heads of the hydra of Greek legend. Below me, before the stage, were lights obscuring cameras and more microphones held by silent reporters.

Behind them was a sea of only angry faces from what I could determine at this vantage point, with the sun and lights in my eyes.

My speech had been placed on the podium. Next to it was a device being used to draw and test the blood of those speaking. I looked out above the crowd as I waved for a temporary stall to gather myself. I had to appear confident in my words if I wanted to sound convincing.

I cleared my throat and stepped up to the podium.

"Good people," I began, "I *know* today you are insecure. You do not know if your *faith* in this administration has been misplaced. You do not know if you can *trust* your neighbor, your coworkers, or the military. You do not know if your *children* are safe at home. I am here to tell you – there is *nothing* to fear! While we have this device here to prove my legitimacy, I will not be using it until I have finished speaking. I wish for you to hear me – hear me with your *doubts* and your *disbelief*. You will hear the *truth*, and it is my hope there is no question as to my identity when I have finished. Then, I will assuage your conscience."

I let the crowd inspect my form. Many of them scrutinized me for any sign of a false humanity. Many continued to yell and call in a fervor rather than listen to my words. I would have to be more direct.

"I wish to assure you, this does not mean war!" I continued with a passionate cry. "There will not be a front here at home! We have defenses across every border and we have allies who have not been compromised! But our *true* shield against this evil is *you*, the American people. You have the passion to protect our *country*, our *values*, without hesitation. You are the great militia spread across this great nation! You would protect your families – and your homes – until your last breath! And *we* are here to protect *you*!" Some of the crowd had rallied. I could see my enthusiasm was moving them.

"*We* are strong! And we will be made the stronger if we remain vigilant *together*! Do *not* let your lives descend into madness – into paranoia! We now have an alien enemy! We do not know where they come from – we do not know how many of them are here, on *our* planet! We do not know their intentions, but we *must* assume they are hostile! And any enemy of this great country *will* be defeated!" Their cries were louder now, and the entire crowd seemed to be one entity. I chose this moment to wave the technician operating the device onto the stage.

"I'm told their blood is clear." I watched my blood be drawn and saw its deep red color. It was thick, and I followed it as it slowly climbed a tube to a small centrifuge.

The crowd saw it too, and a roar of applause erupted in the front row. It traveled towards the back of the gathered bodies as if it were a wave of emotion until it was all I could hear – even up on the stage. This was that singular, powerful moment. They were so flooded with adrenaline, the right combination of words would be the nail in their coffins. This was an opportunity that could not be wasted. As the technician removed the band tightened around my arm, I raised them both to incite the raving mass.

"My fellow Americans – *I* bleed *red*! *We* are *human*! *We* are *human*!" This chant reverberated throughout the city. These words would be remembered by whatever remained here in the United States and other key locations. I gave the crowd a wave, revealing my palms.

They were heavily scarred from the narrative of an old war injury.

Jacqueline was waiting behind the stage with three of my Secret Service members and more of the strange-looking men in black. They led me to a vehicle through yet another hidden path and I was taken outside the capitol.

We drove to an abandoned warehouse and port at Slaughter Beach on the coast of the Atlantic Ocean. We met with Congressman, senators, and other members of the government. Alongside them were many plain looking people, though they surely belonged among us. We had all been trained for this mission, raised in different locations for different roles. I did not know every face and was uneasy.

"This the group heading to Sub Base-3?" asked a man approaching from the dock.

"Careful with your voice!" snapped Jacqueline.

"*Relax*, it's just us now," None of my Secret Service escorts remained – the men in black were members of the Militia Majora. I could see they were relieved to be off duty from their performances convincing the other agents – their shoulders had lost their angled edge.

They appeared almost human, apparently at an intermediate stage in the gene-editing transformation.

Jacqueline quickly gathered herself and began giving us our criteria for the trip ahead. We would travel out on the Atlantic to meet an un-identified *submerged* object and transfer to this vessel for the rest of our journey. The craft would prevent any sonar or other detection device from viewing our movements.

SB-3 was an underwater facility anchored in the Arctic. This was a hub for communications across and outside the planet. World leaders occasionally gathered there to receive new orders directly from the Imprir.

I was excited to be with true comrades.

I consumed what information we had gathered from China. I knew some names and faces involved in the situation in North Korea – through some duplicity, our operatives had developed guilt for their actions overthrowing the previous leadership, and believed they were helping the people by revealing their identities.

The people did not agree.

In China, however, we had a firm grip over the politicians and corporations – like in any democratic country – that were in power. A similar situation had unfolded as in France – before Li Jie came to power, the handler was forced to assume their charges position. This team was infiltrated by a native agent through the use of lust. This detail left many confused.

A relationship between one of us and one of them was not viable in a reproductive sense – they were bipedal, but similarities stopped there.

They were most relative to the reptiles of Earth. Cold-blooded, they stood taller than humans with an elongated form. Three large fingers clawed each hand, and they were covered in scales from the crest on their heads to the tips of their tails. They had eyes like a chameleon, disconcerting when looking in separate directions.

How one would desire such a thing, or develop such a connection, was beyond me. It seemed impossible. As I turned over the idea in my head, Jaqueline approached and informed me it was time to depart.

We boarded two separate fishing vessels that left on staggered launches. They both waited offshore in international waters for the USOs. Once each crew had been collected, the boats returned to port, and the craft traveled North East towards the Arctic Circle.

We were in close quarters and I was eager to arrive at SB-3. Working within the government was a quiet and lonely profession, with brief contact with the citizens or my cohorts. It appears our comrades deeper in the field had become more familiar with their neighbors, and much more...*conversational.*

"Everyone was talking about you, wondering if you were gonna take the test, yeah? It was all they could talk about at the hospital," a man named Thomas said. "They all think they wasted their votes. I just tell them it doesn't matter anyway, their votes don't count. They love that sort of shit. Government conspiracies and all that – wouldn't it be funny if they knew they were mostly true? Just not *their* government."

He seemed to be in the know.

I did not respond, but he was not deterred in the least. Most of the crew on this vessel were members of government, though some, such as Jacqueline, were separated on the other USO. In the event of an accident or the discovery of our mission, one of our leading operatives with the knowledge needed to continue this enterprise would survive. However, this required a trade between the ships' seating charts. Three of our working-class operatives were aboard, and the other two were no less talkative.

I learned Thomas was a genetically altered alien. He had married a single mother and started a family through adoption. This was his true assignment, the purpose of which was to investigate this aspect of the people's lives in order to exploit it. If his feelings were true, it seemed the Imprir had indeed imagined a relationship between our peoples.

He had found that protecting family was the second most powerful motivator in a person's life. He believed it to be the first, but I knew better.

I knew that greed was the great motivator of humankind.

Many would betray their family to satiate this greed, abandoning their relatives and lead lonely lives chasing ink and paper to spend on *things*. The most valuable metals and minerals – materials that could advance human technology *centuries* overnight – were used as baubles or sat in safety deposit boxes collecting dust. Gold, perhaps the most valuable naturally occurring metal, is a symbol of wealth and status. Its true power was in its value as a conductive material and its *abundance*. It was wasted on trinkets when it would be capable of powering magnificent machines, including space vessels.

That was power.

While equally enthusiastic about speaking freely, our other two passengers ended their conversation. Objects were being picked up following us outside the hull. They appeared as blips on the sensors, but those manning the vessel were educating guesses.

"They're giant squid. Nobody believed they were still alive, but once they found them again, started seeing them everywhere."

"Maybe, but I've never seen one down here, interested in our ship."

They followed the ship for many miles, then suddenly veered away. Sensors were not picking up anything but terrain – no fish or other ships. The Sub Base transmitted a signal that kept them away.

We were approaching our destination.

"It's quiet now...," murmured Thomas, ignoring the obvious and constant sounds of the ship interior. The buzzing of the lights, the rumble of the engines, and the beeping of the different apparatuses all around us had become white noise.

The surrounding sea had become pure darkness as we passed under a glacier far above. No matter what world you are from, the lack of definition of any form that is the abyss shakes one to the core of all fears – that there is something lurking in the dark that will always catch you by surprise.

I felt as though disembodied eyes were now on me personally. A faint green glow in the distance appeared. As we approached the illuminated waters, we could then see the ocean floor. Atop it rested Sub Base-3, one of the first permanent structures built to better subvert the free will of the people.

CHAPTER III

Sub Base-3

Arrival had dashed my hopes of seeing many friends. Sergei had not arrived, though including myself, only four of my training unit would join us at SB-3. News had come in of a coup in Brazil and had traveled swiftly over South America to Mexico. All that remained of our resources were that of the United States, Russia, Nigeria, France, and India. That left only five launch codes available – provided there were no more surprises.

It would seem our leadership was overconfident in their ability to conceal our shared intent.

The launch codes were a contingency plan for this very scenario. The alien force would descend from orbit and hold the world hostage – a worldwide nuclear scare more dire than any moment in the Cold War. Seven world leaders would be required to convene at a secure location and initiate the destruction of the Earth.

This and the other Sub Bases were staffed by operatives that officially remained MIA or POWs. They lived solely to serve the aliens on covert missions and cover-ups. Classically in dark suits, hats, and sunglasses, they were incorporated into the men in black narrative, referred to as the Nem-in-black. Their eyes showed no emotion, and they hardly spoke, even among each other.

When they blinked, their eyelids closed from each side, the only remnant of their reptilian heritage.

I knew the Nemitians had a device that could wipe the memory of an individual or alter memories. I also knew that the more conditioning a mind endured, the more of the self that is lost.

I saw Adebowale entering the conference room and stood to greet him. We clasped each other's forearms, bowed, and then placed our heads to the side of one another. We had learned this traditional Militia Majora greeting during instruction.

"How long before the others arrive?"

"We do not know if they have been able to leave their respective capitols. You may have had an easy exit, but...I did not," Adebowale informed me. He had a look on his face I had not seen since we had taken Instruction. It was a look of pure disgust, and I knew where the conversation would be leading.

"Addi...," I said, attempting to calm him down with this pet name.

"I'm sick of these puny beings. These *people*, have no courtesy, no honor, and no intellect! They are practically *animals*, but we let them live on, believing themselves above, top of the evolutionary chain! That they are *important* in the order of this universe! That their lives will serve any meaning!"

"Kaancha...," I said, using our honorary Militia Majora title. It did not incite his professionalism.

Instruction had taught us that humans were lesser beings than these aliens. They thoroughly enjoyed showing us they were stronger, faster, and smarter. They claimed we would be treated as their equal, though I often wondered if they were any more reliable than our own politicians.

Doubts I did not share with even my dearest friend.

Some took their perceived status to heart more than others, Adebowale included, separating himself further from humanity.

"No!" he continued, seeing through my attempts to deter his momentum. "They toil and work their lives away, when they should suffer – payment for their sins! And the politicians we see – the ones whose own ambitions greatly outweigh their care for their own people

– whose personal interest causes the deaths of millions of their population! Power, in the form of paper titles! They would do anything for the promise of a special place – for the fear of the What-Comes-After! Believing the unknown saves the chosen few, and *they* are the worthy!" he cried, his eyes piercing mine with the intensity of his emotions.

"A promise we do not intend to keep," Jaqueline added, always waiting for the right moment to share her presence with those she is observing.

"Nor do they continue to believe we do," I remarked.

Truman signed an agreement with these aliens – the *Nemitians* – in 1953. They were of the same species found in the Roswell crash as confirmed by the Air Force.

The Nemitians had been equally surprised to find life in what they referred to as our "realm."

In exchange for their technologies, they could build a base on American soil in the mountains of New Mexico. It was a mandate of the treaty the aliens not reveal themselves to the global community – specifically, Russia. This resulted in many cover-ups.

America began developing microwave technology and transistors that minimized computing. The Air Force publicly made leaps in jet propulsion and nuclear technologies. When members of the private sector were invited to meet with these beings, their attention turned to our industry.

As the government played a lesser role in their plans, not all who were elected were privy to their existence.

Several Presidents made inquiries into the existence of aliens and UFOs. John F. Kennedy believed they were already working with other world powers, including the Soviet Union – they wished to use the nuclear armament of Cuba as a guarantee of their control over the United States. They assassinated him for this belief in 1963.

In 1979, America learned Russia had joined them at the adult table along with several other countries.

The Nemitians had erected a base on the Moon after the landing in 1969. From this vantage point, they had performed cultural studies for each of the major powers where their infiltration would begin.

Each government had its own methods for this secretive agency with vast intelligence gathering networks acting independently of each other – even under one flag. They trusted no one, least of all those with the freedom of these spies. While boasting of finesse, these agencies acted like blunt instruments.

The Nemitians required surgical precision.

They learned everything they could about the various languages and how they differed by region within the same etymology. These differences were most important if they were to have agents in every part of the world and they engaged much in "small-talk" exercises.

They became familiar with the power of phraseology and how to structure such phrases as to influence the people.

These beings had been making preparations for decades prior, with operatives undermining any administration such that the people wanted new leadership. This made a country ripe for subterfuge.

A clandestine group, both human and Nemitian, was recruiting from the fronts in the Middle-East to rise through the ranks of diplomacy – the Majestic Twelve. Their mission was to ensure our place as world leaders of twelve countries: North Korea, China, India, Japan, Nigeria, France, Germany, Britain, Russia, Mexico, Brazil, and the United States.

Americans had a hunger for more progressive leadership. Satiating this hunger would keep the aliens out of sight, and it was not long before they had infiltrated all of their targets. Each time a country seemed to gain independent sustainability, a maelstrom of trouble would suddenly appear outside the bounds of its influence.

This gave a higher power reason to occupy or invade. Gimmicks substituted for any actual intellectual debate. Uncorroborated gossip shamed into obsolescence any who spoke against such decisions or sympathized with enemies created by the media.

Some simply disappeared.

I was recruited in 1988 by the first President Bush himself. Left for dead after being caught delivering explosives via a case of mangoes in Pakistan, I was a prisoner of war, rescued after my mission was complete through other means. The seven Air Force officers were accompanied by a strange-looking man in dark sunglasses and a black suit, who offered me a new life serving these aliens. I accepted and was introduced to my first Militia handler – an alien and Jaqueline Stone's predecessor.

They had learned to alter their form, appearing human and assuming the lives of those missing in action.

There was a moment when the world was most vulnerable, held hostage by not only one truly *fearless* leader and his unpredictable temperament but also that of several other world powers who had let their governments become so shrouded by dark intentions they no longer controlled their own lives.

All seemed hopeless.

It was at this moment my comrades and I decided the time was right to create the Nemitian vision. Many had their own ideas of how to use their newfound power over this Earth, but our orders came from the Militia Majora.

We succeeded in our first campaign for the position of President in the United States. After taking this office, it was a simple task to replace the cabinet and many other members of influence. Ambassadors were replaced by Generals, decorated for perceived heroisms and with a taste for conflict. Once every arrangement had been made, an Ultima team would be put in place, and it was this team I fought tooth and nail for a position on.

Prior to today, there had been many attempts to expose this conspiracy throughout the decades. They had all been dealt with quickly and efficiently. When my predecessors began, they infiltrated the already paranoia-ridden intelligence networks. While I was taking Instruction, there were several cover-ups using the resources of these agencies.

Complete control remained out of reach. Occasionally, an agent would get too close to the truth, informed by those who were held in confidence by these infiltrators. They had to be dealt with cautiously.

Every incident required revealing some other secret as misinformation to distract the public, one of the primary goals of our subterfuge being to keep doubt alive.

What can one believe if one cannot believe anything at all?

"This is worse than Ellsberg." added Jaquelline.

"Ellsberg, or Vietnam?"

"*Vietnam*," she stated, referencing an operation that had taken place before my recruitment. I had heard the story many times as a deterrent to ensure our loyalty.

1969. The Nemitians had taken no positions in China and had to contain their advancement into Vietnam. They intended this to be the staging ground for our operations in Southeast Asia, but a defector of their own intended to expose this conspiracy. This one had gathered a small group of like-minded comrades and informed subordinates on a compound hidden in the jungle. They were discovered and dispatched using United States helicopters and weapons – flamethrowers enhanced with alien fuels.

The jungle burned for weeks, any evidence lost in the blaze.

While an official report was never filed, a record was made of the requisitions of resources. This reached the desk of Richard Nixon, another President to make inquiries into the UFO phenomenon. The aliens orchestrated the release of the Pentagon Papers to allow time for damage control, and Daniel Ellsberg simply provided a vehicle for their narrative.

"Yes, I've read the reports. They couldn't afford to be exposed – to China or the world. And once they finally took hold of the government–,"

"–censorship of outside influence," Adebowale finished.

Unnamed operatives burned evidence while the world looked at Ellsberg. The Nemitians indicated Nixon in the Watergate Scandal and burdened him with evidence of their own surveillance. The publicity finally allowed the Militia Majora a move towards the White House, installing President Gerald Ford's handler.

"Talking about our old friend, eh?" Mathilde had arrived.

"You served with him?" I asked.

"We arrived together, yes," she replied after glancing at Jaqueline and Adebowale. "Well, he had always been a troublemaker – it was his trademark, ever since his days in the Khano Militia. He should have been so lucky to live so lavishly when he belonged in a cell."

"He did believe he was lucky. But he also believed he was *right*," Jaqueline added, the only other witness. I never missed a tone in a word and knew this was not a conversation she wished to continue.

"Then there was Snowden."

Counterintelligence was especially difficult during the internet age.

While precedent had been made for the treatment of censorship in China, the same cannot be said about the American and other Western fronts in our secret cyber-war of misinformation.

This exposure was difficult to deny *or* illicit doubt. While our true purpose to organize a global surveillance system remained a secret, the paranoia of the people caused such disbelief in the two-party system we had created, we lost our ability to manipulate the apparent will of the people. This resulted in rogue contenders disrupting our design, though individually short-lived.

Personalities were elected over established diplomats. A lack of faith in this government and its investments left the economy in shambles. This did not please the citizens, who were suffering for it. They unknowingly returned the power to the Militia Majora.

It was truly an act of their own free will.

The events in Vietnam presented the most dire case of exposure. Another attempt had been made by an alien operative in Pakistan, though they would not live to share any information.

I would see to that, unknowingly advancing the aliens' agenda.

An agent under the guise of an American Ambassador sympathized with the people he provided charity as part of his cover. He believed they were a product of their environment and their evolution from violent beasts. The people could progress no further because they lived in a world of greed and hungers he believed they created in order to more easily control humanity.

This was not *un*true, for much more easily controlled is a mass with singular obsessions that occupy every waking moment.

They believe we have intelligence – in its truest form, our attention and connection to the arts. They appreciated our music, writing, and great works of the visual that permeated our cultures.

Artists navigated a realm of fantasies that would become technology.

Once that media was controlled, we became less human and more comparable to livestock, corralled into our little corners for a slaughter.

President Reagan's administration had been supplying Afghanistan with weapons through Pakistan, but it was believed they were being deceived. I had been stationed there and was a favorite of my superiors, quickly finding myself performing black ops. I was ordered to deliver the mango case that was to cause the "accident." However, I was captured, my role fulfilled as a distraction from the rigging of the helicopter to crash.

The Taliban camp at which I was kept prisoner was destroyed by a drone strike. I was the only survivor, protected in the basement that had been my cell. After a clean-up team of Air Force officers and the emotionless NIB discovered me, I was offered a new life.

The invasion had already begun. If I refused, I would be returned to America a corpse, my actions in the Ambassador's assassination and others, including one Phil Schneider, revealed to the public. My legacy would be shamed – my only option was to join them.

I accepted.

The camp was burned along with the poppy fields the Taliban harvested to fund their campaigns of terror. I was issued an enhanced flamethrower, much like those used in Vietnam, and aided in yet another cover-up.

My mind returning to the present, I suddenly knew I was not alone. I had wandered into the mess hall, and sitting at a table was a man in overalls and a green plaid shirt. He was wearing a dusty hat.

"Oh, Mr. President, how *honored* I am," he said sarcastically, quickly returning to his sandwich.

"If I might ask your position, sir?"

"Oh I just run a farm of genetically enhanced produce to keep 'em docile," he said, never pausing his chewing. "I'm just a nobody round here, ready to get back home."

"What's back home?" I wanted to move the conversation along rather than be consumed by my own thoughts about the past.

"Nothin' special – but a nothin' I know," he replied, tapping his finger several times. "You don't remember me from Instruction, do you?"

I did not, but it became apparent he had been in the same class as Mathilde, Sergei, and I. After the first three Trials, he must have dropped into a lower class because of his scores. This would be the group that would infiltrate the working class. Their roles were that of the conspiracy nuts, pushing disinformation through social media posts and attending mock protests.

It was a position he seemed content with.

Luckily, Mathilde entered the room, looking for a snack. She had remembered our companion – a Mr. Walton – and that seemed to alter his disposition. It became apparent he too was an alien, his form altered to that of the life he had assumed. After sitting and eating together for one hour, Mathilde and I returned to the main conference room.

Adebowale sat with his naked feet on the conference table.

It was becoming less empty as another ship arrived. I noted several crew members but did not see their superior. I spoke with an engineer that revealed to me the Captain had been lost in an attack. This ship was coming from India, and Rajesh Bakshi, then President of India, had taken command of the vessel and had navigated to SB-3 with limited available sensors.

They had lost communications and brought no word from Sergei's sub.

I found Rajesh as he was being treated by a medic. There were medical facilities on board, but he was stubborn. He wished to oversee the crew count.

"11 unaccounted for," he said as I approached. "What does that make it today? Two – three hundred?"

"I stopped counting in the last war."

"Smart man." I could see in his eyes that the deaths were weighing heavily on him, though his demeanor would not betray him.

"Raj...it wasn't your fault," I consoled him. "You took control of the ship in a dire situation and saved more than you should have, with that lame arm." I added, jesting an injury of my own.

He laughed and then a grave expression held his features as he informed us of another loss – our comrade Chirrak. He had taken the form and name of Rudra Bhatt, but his true name should be used after his passing. Adebowale joined us, though Mathilde – *Turrina*, rather – took to the porthole to stare into the ocean.

"Remember when we were taking Instruction, he would sneak away to climb the birch trees outside?"

"The branches would never support him, and he spent most of Instruction in the infirmary!" he added with a chuckle.

"Of course – how else would he have had so much time with Turrina."

It was during Instruction Turrina and Chirrak first met in the infirmary. She wanted nothing to do with him, but was forced to be at his bedside for two weeks. In that time, the things that had once annoyed her charmed her. What she had labeled his faults became his uniqueness. She knew they would be together, but she would never tell him that.

Now, she would never get that chance.

We continued to reminisce as we awaited news from the others. The only one missing was Sergei. It would be a lie to say I was not getting worried about him. As I moved away to console Turrina, gazing through the view-port into the cold ocean, I felt a chill through my spine.

Some believed that meant a spirit was passing through you. I could not help but feel haunted by those already lost to what had become our crusade. I then heard the beeping of an incoming communique.

It was Sergei.

"What's he saying?" I asked fervently as I approached the communications panel.

Rajesh looked up from the panel, then back down at the controls. He raised the volume and I could hear the anxiety in Sergei's voice.

His ship was under attack.

"- ships, three subs," his voice cut in. "They followed us from Russia outside of radar in order to find the base! I took the pills, passed the test. They convinced me I was safe butzzzzzzkkrrrsssssss."

Either the ship was attacked or the feed was being interrupted. We did not know how long we had before the enemy was here and – in our current position – there were few defenses.

The Sub Bases were, in fact, of alien design. They were anchored to the ocean floor in the Arctic and Antarctic Oceans. Preparations would be made to release our anchor, but this vessel supported a great weight. It would not be easy to dislodge it from the rock and ice after it had rested here for several decades.

The other members of my team were distributing the many tasks now needed to be completed. I returned to the window and gazed out at the ocean beyond. I stood there and imagined what was to come. Everything had gone astray. The people knew *more* than they should, much sooner than we had intended.

Our benefactors would need to revise their scheme.

"What is that?" came the voice of Turrina, with her finger pressed against the glass.

Returning my eyes to the view-port, I saw a shape materializing in the distance. It was growing larger – moving towards our position – but no clearer.

"Could it be a creature?" Rajesh asked as he and Adebowale joined us.

We were all gathered near the wall as the shadow grew larger still. More appeared above and on either side. Soon the first shadow became distinct – it was the USO carrying Sergei and his crew, the smaller objects their pursuers.

Sergei's vessel was moving erratically – it was apparent he had lost control. The heavy vessel was floating freely through the water, propelled only by its momentum and alien propulsion. These craft could travel *much* faster than human naval vessels.

Alarms began blaring and sirens flashed all over the ship – Sergei was on a collision course.

By the red lights flashing like strobes to illuminate the room, I could see a frenzy of motion as the many bodies ran for the doors on either side of the conference table opposite the porthole. This was not the control room, nor was it the most fortified position on this ship.

With the room cleared, we could seal the doors to prevent flooding any other compartments should the hull become compromised. Even those in civilian roles were trained soldiers, but I could see only panicked faces and furrowed brows.

I slowly made my way across the room, watching the others rush out on either side. As I looked over my shoulder, I saw Adebowale and Rajesh trying to pull Turrina away from the window. She was crying.

"I don't want to leave! Not without my Chirrak!" she called out, in a deep sadness apparent in her heavy low heaving and bellowing. "Leave me! I can't!" Raj and Addi were attempting to raise her off the ground. She would not budge, and the ships were approaching fast. Addi and Raj argued over whether to leave her, and through the porthole, I could see there was not much time.

I marched over to rally them to escape what could be a catastrophe.

"Come! All of you, we can argue and wallow on the other side of those doors! We are in *command* here until we reach Stage Three! Look behind you!" Raj and Addi turned to see distorted faces – a mix of each of our crews. They would not close the doors without us.

They looked at each other and headed for the doors. Turrina remained crying on the floor.

"Mathilde! *Turrina*!" I called, doing everything I could to release her from her trance. She shook loose and crawled to the window.

"Chiki... Chiki...," she whimpered.

I turned and ran for the doors.

The moment I reached the other side, Addi nodded to the engineers at each position and signaled for them to seal the doors. Loud clanks rang out, and a hiss erupted from the wall as the compartments pressurized themselves individually. Turrina would be trapped, but if the hull held, she would have enough air to last until our defense was complete.

I braced myself, but it was not enough. The last thing I remembered was seeing Turrina collapsed on the floor, letting out one last cry. The base could not stabilize itself – the force of the impact must have loosened an anchor.

I slammed my head against the wall in the quaking and blacked out.

CHAPTER IV

Proudhome

I was surrounded by shadowed forms.

I felt as though I was paralyzed and a great weight stopped my lungs from drawing breath. These forms were alive – chittering amongst one another. As I attempted to reach out for help, I felt an evil presence nagging at my soul. I strained to face this demon, attempting to turn to the left. Suddenly, I was sinking, the forms becoming elongated. The darkness gave way to light and the evil emotions swelling in my chest were replaced with tranquility.

I lost my identity.

"They have potential," came a voice I recognized as a paternal figure from the void.

We are not your father, came many voices in my mind.

I found myself bound in a hot basement. A stream of light dimly lit the room before the door was opened. I was blinded as light and heat filled the space. The fragrance of burning flesh assaulted my nostrils.

"You will do great things," came the voice of the rescuing officers in unison.

Still you do not recognize us? inquired the voices.

Memories continued to flash through my mind with no particular chronology. The only common factor between them was that they were the memories of an individual – to whom they belonged, I could not say.

My identity remained lost in this growing cacophony of visions.

Don't you recognize my *voice?* I heard within the thunderous chorus.

This voice was separate from the memories, though I could hear it under every other word. It felt familiar, but I could not identify its source from any of the moments replaying in my mind. I saw this life pass before me several times – each cycle brought more memories. Entire lifetimes were playing out above and below each other. The volume was tremendous. Before I could identify any source, I was returned to the void abruptly.

I persisted in this sudden silence, the space I now occupied infinite in every direction. The *self* remained lost to me as I strained for any voice. The void was not empty – it was filled with light burning away at my soul.

An eternity was passing – and yet no time passed at all.

A great fear took hold as I realized the utter loneliness that pervaded wherever this was. Over the silence, I could hear the blood flowing through my ears. The *only* sound, it was deafening. I remained in this agonizing pain until that unique voice returned.

You are living forward...you cannot yet see. But if you can hear us, it must be soon. You are realizing the dream.

It was *my* voice. A stream of memories again passed through me, fluid and defined, the last of which was Turrina kneeling at the view port of the conference room, crying. I watched as the wall exploded and water began rushing in.

I shut my eyes tightly.

* * *

The floor was at an incline.

I awoke disoriented from a deep slumber. The details of my experience faded as I found myself under a crew member who was not as fortunate – I could see their neck had broken.

As I shoved the body aside to make my way to the nearest exit, I recognized Mr. Walton.

I had the distinct desire to turn left and head for the control room. I fought the urge and turned right, heading for the sealed doors. As I rounded the corner, I saw Addi and Raj were already standing at the view panel with grim looks on their faces. The panel showed only static, and I assumed there was an issue with the camera inside the room.

"Mathilde-," I began.

"*Turrina*," Raj hissed back. I remained silent as I approached.

The hull had been ripped apart and allowed the sea to flood inside. One minute had passed, and the room was not completely underwater. In the pool covering almost half the room, a large oily area sat above the water with a dark shape floating in the center. The stain grew as more water bubbled in from the hull breach. We watched, holding our breath, as the shape was dragged down in the turmoil.

"Turrina...," I finally exhaled as Addi grabbed Raj and I by our shoulders and pulled us back from the panel.

We were still under attack.

We made our way down the hall to command. Nothing out-of-this-world was used in the Sub Bases – even the anti-gravity system was of the native technologies. With alien knowledge, we now knew how to more efficiently use nuclear power sources. However, that meant using human weapons, which had limited destructive power underwater.

"Torpedoes armed," a crewmember called out from the defense controls. "I'm ready to blast them, Kaancha."

"Do we know where they're from?" I asked, ignoring the honorary title.

"They followed from Russia, sir."

"Blast them."

A single salvo of torpedoes left their launchers. The subs had not turned around, though it was surely Sergei's ship that had crashed into the base. His pursuers intended to fight.

They were not hit with our first volley and had begun their own assault. The ship shook slightly, but there was only minor damage. I moved to view the ship's status display, relaying that the conference room was now completely flooded. There were several spots on the outer shell threatening to rupture. I began calculating the odds of the worst-case scenario – an anchor had been hit and we could not get loose. The ship quaked more violently. I fell to one knee and checked the panel.

Worst case scenario.

"This is no longer a flight situation!" There was movement all about as the navigation systems were abandoned. All eyes and all hands on deck would be needed to defend the base, and I'm sure we had lost more crew to the last blast.

"Let me hit give 'em another one, sir!" As I looked upon them, I was again living in a memory. An image flashed across my psyche of some past comrade. The form before me was visibly fit, muscles filling her Militia Majora uniform. Her hair was as short as Jaqueline Stone's, although a deep auburn.

["Save our shots for a sure thing, Alpha Yekk,"] I replied with the most knowing look I could muster, not realizing I spoke in an alien tongue.

We needed our subordinates to believe we had some plan. We needed to get to the anchors and see if they could be disconnected manually, but the fastest route through the conference room was unavailable. As I tried to imagine a map of the base in my head, a hand grabbed my shoulder.

"It's Major Angela Johnson. Here, anyway – sir, but looks like I've been promoted to saving your ass."

"Yes, of course..." I was distracted by my delusion and an attempt to develop a plan of action.

I noticed Raj and Addi staring at me from the inventory manifest, searching for anything we could use in the upcoming battle. They quickly lowered their gaze.

"Sir, the anchors."

"I know, but the conference room is flooded and -,".

"*And* I've been stationed here the past *four* years. There are many ways around this base, sir, if you top brass types aren't afraid to get a little dirty," She glanced over at my fellow world leaders. "Follow me."

We ran into the hallway, the Major in the lead with Raj, Addi, and the remaining crew, following close behind. Rounding a corner, we passed the communications room as smoke was billowing out of the doorway. Johnson stopped for an instant, hardly breaking stride as she continued on.

"Shit...," I heard her say under her breath.

"What's wrong?"

"There is a maintenance corridor that runs from Comms to Mess, and from there we can – could have – reached the anchor housings."

"Shit, indeed," I replied. My companion glanced back in my direction and I could see the beginnings of a smile.

"Yes, *indeed*, sir. We'll have to go through the cargo bay, where the housings are located. Just another corridor and – "

Another explosion rocked the base, throwing us all to the floor or a wall. This blast had come from directly ahead – from inside the craft.

"What was that?!" Addi yelled over a new wailing of alarms.

Major Johnson was already up, pistol in hand. She moved about, looking down each corridor before returning to help me up. As I stood, a pain arose in my leg. It was broken.

"You're not going to like this, sir." She eased me back down to the floor. "It came from the hangar. It seems we've been boarded."

Johnson and I took up positions on either side of the corridor. We waited to hear the hiss of decompression as the sub docked to the base, depositing its occupants in our line of fire. Smoke filled the air, sparks flew, and I was quite jarred from the explosion.

Currently, the Major held the only weapons in the room.

She tossed me her sidearm and braced herself on the wall, drawing her preferred weapon, an outdated personal railgun.

I recognized the insignia, similar to that of the Militia Majora, emblazoned on a blue cloth hanging from the barrel now wrapped around her forearm.

"I've seen this before!" I called over the many sounds of the attack and alarms. "Where did you get this weapon?"

She blinked once in disbelief.

"It was my generational's, and hers before her! We were named after the first! All served!" came her reply. Now it was I who was in disbelief, assuming I had met her mother in the field or during Instruction. I knew this name – *Yekk* – though why I would recognize the body before me, I did not know.

"You took the same form, it seems!" There had been many recruits and volunteers to join the cause *after* I was in Instruction. Some may have come from Nemi and it was not impossible the aliens would reuse identities and their image.

We shared one last glance – one last look at a friendly face – as light peaked through the opening doors. When the first figure showed their boots, we let out a sigh of relief as we recognized the present insignia of the Militia Majora.

"We are all born Militia!" the first called out to what were only hazy figures in the dark, their weapon ready.

"To serve the Imprir – to die for Nemi!" Johnson responded.

The soldiers, who had been slowly advancing on our position, now rushed past us and down either side of the T-section. Some helped the rest of our party board the ship, followed by an officer who extended an arm to the Major and myself.

"Mr. President, it seems you may need to replace your Secret Service." She leaned in to whisper quietly, "They'll let anyone in here these days." She then dropped several earpieces on the floor.

Colonel Yvonne Stromsky. She was head of all Militia Majora operations on Earth – head of the current Majestic Twelve. She had made some aesthetic changes, ever changing her cover, heavily involved in what would be marked black ops.

"Colonel, you have our gratitude." After a few awkward moments of silence, I turned and headed down the corridor.

"Not gonna introduce me to your new friend?" Turning to Johnson, she chided, "Made you burn any *evidence* yet?"

"Colonel!" I snapped. She glanced at me before cackling maniacally.

I recalled her having a certain sarcastic attitude for an officer – bordering on insubordinate – but she was extravagant in her present manic behavior.

"I could tell! You've been thinking about *him*, huh? I remembered! I'm sure the other Kaancha are around here somewhere too..."

Without turning to me, she articulated her next words:

"It's no coincidence we're all together again, you know. *All* the players."

* * *

It had been a long day. A lot of excitement. Maybe that's why the Colonel had to repeat herself several times.

"The traitor is...*alive*," I finally managed.

"He never *died*, his death was a ruse, so he would have fifty years to tear down the veil we've put up. At least that's going to be what they tell you."

I did not acknowledge her last words, as I was suddenly overcome by the smell of the burning flesh. A new memory played in my mind of a dark basement. The floor was scorched under several bodies I could not identify. It was subtle, but there was a shimmer to the alien blood presently obscuring their faces.

My world was shaking. I felt ill and my vision was blurred. I took a deep breath and tightened all the muscles throughout my body to start my blood flowing. After a few moments, this vertigo subsided, and I stood to announce my intentions. However, a sharp pain in my side only allowed me a quick gasp before I fell back to the floor. The sight of hands, slick with dark red liquid, did not disturb me, but the feeling was leaving my extremities.

The pangs of guilt I felt from this memory faded from my senses.

"My blood," I informed my two companions in the corridor. This made the Colonel laugh.

"We'll get you some help, sir," said the Major as she left us to find any remaining staff – medical or otherwise. Yvonne followed her exit.

As soon as she turned the corner, the Colonel leaned in close to whisper in my ear.

"I thought I was the only one left, then I remembered! Just the two of you now! But not if you don't get up!" She lifted me up to support my weight with her shoulder, causing me to wince. I was led back down the hallway to the hangar bay where a ship waited. Not an escape pod or a submarine, but a proper vehicle.

I was placed on a bench in the cargo area of the ship. It was not a spacious vessel, usually carrying only a few soldiers aboard. It was obvious they were not sent to recover any survivors, but it was my luck that I was important enough to be saved.

Though my mind was racing, my body was exhausted. I settled as best I could on the cold metal surface and closed my eyes. I dreamed of sleep, though it would be keen to get rest of any kind before I was needed. This situation was far from over, and the next move far from decided.

I lay there, thinking of the different paths we could take.

The Imprir could decide that this secret invasion was no longer worth the effort. If there was to be a war, it would be a quick one. The defenders would be wiped out and their great cities obliterated. Any who remained would be enslaved. Ultimately, they would not survive.

I would have little to do with that.

The second option was that we continue this farce and attempt to salvage the faith of the people. We would support any revolt in China and Russia, then find a fringe group to build a station in new territories. Acts America has perpetrated before – feigning support for the oppressed and inserting a figurehead that we control. This option would ask much of me, but I intended to perform my duties as I always had – without question.

While trying to divine a third path, I was startled by the sudden drumming vibrations of the ship's engine. I had rarely been around such technology and I failed to recognize the EM-field generator giving the ship its artificial gravity and a weightless feeling all at once. I slowly found my way to the door, hand at my side.

"Ah, I would have called for you if we weren't in such a rush, *Mr. President*," I heard as I passed through the open doorway. The Colonel kept her back to me, but several heads gave a quick turn at the derision in her voice. She was now wearing a pilot's bomber jacket with the collar up. The only personal item worn by any of the crew, this truly was her vessel.

I would allow her this slight.

I saw Johnson was also aboard, as well as a few other survivors. Missing were my two friends, though the cabin was cramped and hot with bodies as it was. The water tugged at the ship as we ascended. As I looked out the view port, I saw several other ships leaving key support sections of the underwater base and realized their intentions.

The blasts blew the base from its anchors, and it sank slowly. The massive shape disappeared in the turbulent waters now wrapping around it. Cracking, the rock and ice that sat above the base broke off in large chunks, attempting to rise against the glacier above. This would bury any evidence of the base and its technologies. There was a solemn look on all that viewed the destruction.

This was a great setback to decades of espionage and infiltration.

As we cleared the waters and rose slowly into the clouds, I looked below to see a complete neighborhood sink into the sea. There had been a small fishing community here, with a short landing strip we had all used to travel back and forth between our respective countries and the base. I had never walked through the town, but I had seen the sign at the end of the runway each time.

PROUDHOME

POPULATION: 343

Three hundred and forty-three men, women, and children. Three hundred and forty-three lives. Three hundred and forty-three souls.

Now they were three hundred and forty-three units of collateral damage.

A minor loss of life should not have any effect on a seasoned soldier and this was not an American outpost – it belonged to no country. I returned to the cargo area, staring for several moments at the wall before slamming my fist into the cold metal and falling to the floor. I attempted to rest, wishing to be ready and willing to discourse with our superiors.

I had many thoughts before drifting off into sleep, the last of which mirrored my earlier worry:

A minor loss of life...Shouldn't be affecting me...

CHAPTER V

The Moon

None would be excited about the upcoming meeting with our *true* masters. In the long journey, though a few hours in the alien craft, the atmosphere in the cabin had become heated. The ventilation was not meant to filter air for so many lungs.

Those genetically able were sweating.

The heat was only exaggerated by the situation at hand. I could draw many correlations between what I had learned of Nemitian history during Instruction and the past few decades. The Imprir had themselves replaced what they called the "Old Government." It was a coup much like what now plagued them on Earth.

The Old Government was an uneasy alliance of the three factions that previously held control over the territories across the two moons they inhabited. Long ago, in an age where truth has been lost to myth, their people were said to have been deposited on a habitable moon orbiting a gas giant in their home system - Nemesis. Several claims were made, resulting in the first wars to spill blood on this virgin world called Nemi.

Those that survived spread themselves like a virus to every corner of the proverbial Eden under the twin suns of the system.

Technology became industry, and that industry required more than their world could provide. Nemesis had two more satellites that luckily never intersected the orbital plane of Nemi. Neither had an atmosphere, but when they became capable of space travel, every company saw potential in the seemingly endless supply of raw material. One such company – Khano Industries – was the first to develop a vehicle for space flight.

Their focus was a reliable, reusable transport made up of the metals found naturally on the nearest moon. One of the two world powers – five hundred years later, which one remains unknown – contracted them to build a base on the moon designed to stage troops for some future maneuver. Using the capital from this contract, Khano Industries built a factory and workers colony with a population of ten thousand before the contract expired a decade later.

Over that decade, the company moved all of their physical holdings and secure servers to warehouses and offices on the moon. Once wholly independent of their reliance on any government, Adunaan Khano claimed the moon as his own State, their own body to govern.

It was dubbed Khano Alpha.

While the corporation could tax the export of almost all the world's material infrastructure, those living in the workers' colonies were now slave citizens of this newly created power.

Ensuring the workers' cooperation and enforcing the new Rule of Law were Khano Industries secret weapons. The true breakthroughs they were making in their unchecked facilities were in artificial intelligence. The same government who afforded them the properties had asked for a controllable military force with a desire to dominate all of Nemi. These androids, resembling the soldiers they would fight, were Khano's guarantee of their hold over the moon and a newfound power source – the He-3 molecule used in controlled and waste-less fission.

It also powered their new rad weapon technology.

A truce was made between the two countries of Nemi, soon to meet in regular summits with Adunaan. A treaty was signed and no one power would be provided for in lifeless soldiers or weapons more than the other. Nemi saw peace under the Old Government.

With no military applications, the AI substituted worker wages with free labor. An inexhaustible workforce progressed the infrastructure of both powers while the citizens suffered, unable to find employment.

Greed led to espionage, and deceit led to a civil war.

The terrestrial States deployed these artificial workers-turned-soldiers. No lives were lost until an explosion at the Khano factory producing the androids obliterated the AI network that controlled all the living weapons on all sides. Having become destitute, the displaced working class refused to fight for their leaders.

The people revolted, disbanding these two governments and inciting others to do the same. The vacuum of leadership left the communities in anarchy, allowing for other corporations to move in under the guise of stability and careers for all. They would impose their own laws through subsidiary companies in law enforcement – their own personal Militias.

With complete autonomy, these new countries taught competing histories and false evidence to their populations to make an enemy of neighboring states. Each community had a newfound – albeit mis-guided – sense of national patriotism and local unity. This provided plenty of volunteers to the Minor Militias, positions that promised stable employment during and after the impending conflicts.

These corporations – media conglomerates, construction firms, and weapons manufacturers – decided in was in all of their best interests to unify. Each appointed a Head CEO to the Imprir. These were to be world leaders and included Aduunan Khano, the first High Inquisitor, leading this one-world government.

Some believed these were his plans all along.

The Militia Majora was created by the Imprir to be impartial, accept-ing applicants from any of the Minor Militias. However, those first drafted were completely loyal to Khano Industries. Years later, when the Imprir made a proclamation disbanding all Minor Militias, the Militia Majora absorbed many of their ranks.

This gave Khano Industries a controlling majority in this combined military force.

Aduunan drove the policy changes until his death at the hands of his generational successor, Hi'Leng Khano. Traditionally, positions in Nemitian society were passed down through familial generations and she was given the option of leading the Militia Majora *or* retaining Khano Industries.

To that effect, Hi'Leng relinquished ownership of her father's company, claiming herself Named Superior of Law and ordered the Militia Majora pledge allegiance to the Imprir. It was a position taken by any who should defeat their former.

The title would remain in the Khano generational line to this day.

The Imprir believed there should be a separation between the military and policy making corporations, so Adunaan Khano's holdings were auctioned under two conditions: those employed by Khano Industries kept their previous positions and the companies owned by the corporation would not be dismantled.

The Imprir agreed and the position of High Inquisitor became that of an elected official, chosen by the other members.

After heated deliberations, Khano Industries diverse portfolio of technology patents, weapon schematics, and manufacturing plants, went to the Mirrit Corporation. Their industry now focused on infrastructure and transport economy – including this ship now gliding across the surface of Earth's moon.

We crossed the boundary of light separating the visible and dark hemispheres of the Moon. From the planet, the satellite would appear this night as a waning half. Though we were now obscured in the deep ocean of black, the cloudy blue pearl that was home to humanity showed vividly my world, now so small.

I suddenly felt myself empathizing with those first of humankind to share this view. I could see the world in its entirety and realized how small I truly was – how miniscule the entirety of my life would be.

"We think the initial leak came from another Sub Base – the encryption is only used on *our* most secure servers. There are few with access to this information on-planet," confirmed Colonel Stromsky.

"Or off-planet," added a communications officer from SB-3.

"Perhaps...," responded the Colonel, keeping quiet. There had not been a chance to off-load any of the passengers, including those considered non-essential to the upcoming conference with the Imprir.

"Is there anything else you'd like to share?" I inquired of her.

She turned to me with a criminal smile on her face and an intense look in her eyes. "Oh, so much more," she said, turning away. "But I don't have the privilege of sharing *that* with you."

We decelerated, though nothing was visible within the range of the forward lights. Suddenly, within a few inches of the ship, the air appeared to split open. A blade of light cut through the darkness – a projected mirage.

The ship was dragged into the hangar, gently resting on the hard metal floor next to another ship that had arrived earlier. As we exited the craft, the gates remained open, locking their position with a heavy and deep *CLANG*! A blue glow tinted the exterior darkness, marking the barrier retaining our air supply.

Upon exiting the craft, we were separated into two groups. One was composed of myself, the Major, and Colonel Stromsky's crew. The other was led by the Colonel and the stranger to the opposite side of the hangar. This group was composed of the comms officers and others rescued from SB-3.

The Colonel read the names listed on a LightPad, a device that projected a reactive screen. I heard one name – Purseis Comaan – followed by a soldier stepping forward.

The rest would be screened to confirm their identities. The alien in black raised a fist and revealed a wristband device. It flashed a particular pattern of light directly at the retina. Aliens in civilian cover were conditioned in this way to truly believe they were human. These false identities could be decommissioned by reversing the pattern they were originally conditioned with.

I was surprised when the Colonel approached and raised the device to my eye line – suddenly, a lifetime of memories were altered as a second perspective and personality manifested itself.

It had always been present, hidden behind my vision and watching as I performed my duties for the Majestic 12. It had seen every life I had taken to continue this enterprise. I squeezed my skull and collapsed as my life on Nemi streaked through my thoughts. What had been out of reach was my true identity – Kriss Methius.

Nemitian.

This was the great pain of our metamorphosis.

In fact, each of my comrades posing as world leaders were truly alien. Our handlers ensured that our conditioning remained active after the incident in Vietnam showed its necessity.

After the betrayal of my generational former.

The final memory before my conditioning was of an operation in Afghanistan. I had not completed the transformation and was serving undercover as one of the NIB, assisting the Air Force to reclaim a captured operative – one Oscar Orwell. Unfortunately, his body was found within a collapsed basement structure.

I needed an identity for my new life as a human and circumstance had provided me with a man that would have returned home a hero.

His body was left behind, burned along with the fields of food the village had grown for harvest. Although the Taliban controlled this area, they preyed on these defenseless communities. They hid behind the innocent, knowingly making them a target.

The Major assisted me in rising off of my knees as the torment in my mind subsided. I knew that my earlier injuries would soon heal as well due to my true physiology, though my side continued to ache. I watched now as another collapsed in this pain.

"Good to have you back, Kaancha," she said.

There could be no risk to The Imprir from any native life forms. While the decontamination would rid us of any lingering bacteria outside of the body, those that exist within and other compromising organisms would be considered a threat.

The process did not discriminate those that are necessary for *human* bodily function.

We disrobed and passed through the gaseous chamber, safe for us to inhale. In the adjoining room awaited sterile clothing for us to wear until we reached our quarters where proper attire awaited us. Before anyone had thought to speak, a commotion arose from within the hangar after several shots were suddenly fired.

Only one of those screened no longer stood near the wall and I now recognized them as Phalae Omega Purseis. Colonel Stromsky had provided him with a standard-issue rad pistol of his own. The strength of the blast was controllable – ranging from a stun setting to a near explosive reaction. At their most destructive capacity, these weapons disintegrated any organic material their radioactive charge came into contact with. It causes the cells to undergo a violent chemical reaction and spreads like a virus. It was a quick death and left only fragments of bone to mark the life that had stood there.

The Colonel adjusted the charge of her blast and provided the first shots, executing two of the humans before the others panicked.

Four to go.

Purseis fired one shot as the comms officer – a Travis, I believe – pushed past them. He was trying to get through the decontamination chamber. The other soldiers ignored him as they disintegrated those frozen in terror. Only Travis remained.

The door to decon on the hangar side was open. As he clawed at the exit, surely begging for help through the all-things-proof wall before us, the Colonel entered and closed the first door. This activated the system, releasing a stream of liquids and gases whose sole purpose was to corrode away the cellular structure of any organic material native to Earth, of which Travis was one hundred percent. While some turned away, the Major and I watched as his blood boiled under his skin.

His features distorted – the lips and eyelids shredded as his hair fell off his skull with patches of scalp. The eyes liquefied as his face pressed deeper against the glass. Blood stained his fingers and soaked his uniform as he reactively tore at the pain on his face before losing consciousness. The body then fell to the floor in furthering states of decay.

The form was enveloped in a dark mist as blood evaporated and mixed with the gases already filling the chamber. A bell rang as if to mark his passing before the doors opened.

The mist cleared behind the doorway. Colonel Stromsky stood proudly with a grin on her face. It was a much more painful and prolonged death than by blaster.

One should not wish it on their worst enemy and I found it cruel and unusual of the Colonel.

"The look is always the same." She used her boot to cock the smoldering head towards us. Behind her, Purseis placed the bodies by the gate. I could feel the Major's eyes on me, but when I turned to face her, she could not meet my gaze.

"As I expected, not much of a reaction from this one," The Colonel continued, as if providing a report on my behavior. Addressing me directly, she asked, "Or is the scene truly that familiar to you?"

"I didn't kill *any* of these men!"

She smirked as she passed me by.

"You used to call them animals – and worse."

Purseis disposed of the remains and then joined us in the antechamber, as we all prepared to enter the base together.

The base was mostly abandoned. It had served as an observation post before any Nemitians returned to the planet after first contact. It possessed mining and refinement equipment for the mining of He-3, currently out of commission. The base also shared similarities with the colony Adunaan Khano had built on Khano Alpha. This included checkpoints surrounding the primary structure where AirMo devices could be stored to ease movement around the base – metal orbs one could remove from the wall. With the press of a button, the scooter was projected in a solid-light state similar to the fake barriers; however, there were not enough of these anti-gravity scooters for each of us

After passing down two corridors that took us deeper into the base, our party had depleted itself of the soldiers – excluding the Colonel – who now guarded either side of the preceding passageways. I had hoped the Major would remain by my side, but it seemed even the Imprir is on high alert with the news of a saboteur in our midst.

When we entered the conference room, the Colonel left us alone to await our debriefing. As she entered the adjoining office, I took note that the desk lamp was on, casting a familiar shadow on the wall. Only one character among those high-ranking officials carried such flamboyant relics as the Staff of the Speaker defined by the dark shape. The light reflected off the shimmering scale wrapped around the artefact.

It was the unmistakable gleam of the crystal skull capping this shadow that marked the presence of the High Inquisitor.

The purpose of High Inquisitor was to provide an impartial party should a case challenge the Imprir. It was extremely rare that a company challenged a major conglomerate unless an inquiry was made at the personal request of the Named Commander. Thus, this position became that of a figurehead with no oversight. It was a position to be desired by only certain personalities, some of which would use their agency to indulge in any manner of vices and secret projects.

Attendants provided food and water while we waited. We were all suspects, but there was not much discussion to be had with the civilians. The attendants quietly observed us to ensure we did not corroborate, as this conspiracy was surely a group effort.

There appeared to be an argument and Colonel Stromsky threw open the door and shot me a wild scowl. After her loud exit, each of my temporary companions was then taken into the office one at a time until only I remained.

The High Inquisitor was saving me for last.

The last of the SB-3 crew left the office and gave me a nervous glance before continuing past the large table in the center of the room. I stood and rounded the table, not wanting to give the Imprir the opportunity to call upon me, ordering my presence. I preferred to face this questioning and was ready to prove my innocence. I stopped before the door to collect myself, but as I reached out my hand, it swung open. The sudden sight looming over me that was Hi'Ran Akeht sent a shiver down my spine.

"Where is Jaqueline Stone?" he inquired.

CHAPTER VI

High Inquisitor

It had been some time since I had seen one of my people in their true form. While no stranger, the High Inquisitor was himself unique even among my kind.

To visit other worlds, we required apparatus to regulate our body temperature. I could see the shape of such a device obscured by the flowing pleated purple robe signifying the rank of an Imprir. Draped over his shoulders was a shining emerald shawl. The light from the room beyond was obscured by the wide collar around his neck – this cast a ghastly glow about the High Inquisitor.

My eyes met his as I took in his full features. He loomed over me as any Nemitian would over a human. His eyelids circled one beaded black eye – the other mechanical. It was a remnant of a war long past and now boasted a blue hue that would blind any who looked directly into it. Male Nemitians usually had two horns cresting their nose but the High Inquisitor's had been removed with half of his face.

He was a cyborg, mechanized over half of his form.

This was not the only thing separating the High Inquisitor's appearance from that of the common. Most of my species had a crest on their head – a fin resting above the skull. Some were born with a birth defect that suppressed this gene. This manifested several recessive expressions that cause the skin to remain scale-less. The tympanum, soft tissue that covered the ears, was enlarged. Fins sprouted from the neck and the High Inquisitor's were quivering with excitement.

"Anything you wish to state before we begin?" His living eye searched mine. I tried to remain calm, but he could see I was intimidated. He pushed past me and sat at the center of the table.

"Come," he bade me to the opposite side. "Let us begin."

As I sat to face him, I felt how small I truly was in this room. The dimensions of Sub Base-3 were constructed to appear human as a precaution. On this base, the size of the door, the chairs, and the height of the table seemed of goliath proportions in this body.

I fought back my vertigo.

Several screens appeared above devices placed in front of the chairs on either side of him. I reigned in my anxiety such that I recognized other Imprir appearing alongside the High Inquisitor. These holograms were another technology reserved for operations off-world.

Behind each form was a roboticized orb, floating at eye level. These were their sprites – personal assistant AI that pervaded Nemitian life. They projected personalized avatars I could not see, as a hologram cannot project another active hologram.

Most would feel honored by the presence, though simulated, of the most powerful figures of our realm. It was rare they had reason to gather unless the Imprir had been dishonored.

After a suffocating silence, the High Inquisitor turned to face me.

"A plot has been discovered to uproot our hold over Earth." He tapped his three fingers before him and glared down at my diminutive figure. "Those unseen enemies within and without would see us at war with the people we have spent so long subjugating – would see them wasted away! Our laborious livestock reduced to but a few *museum* pieces." He paused before again asking the question that seemed to drive this manhunt.

"Where is Jaqueline Stone?"

I recounted my every action after the discovery of this plot to expose our interests. I described my extraction to SB-3 and any pertinent exchanges I took part in. I could not provide them with the where-abouts of Jaqueline Stone once we separated upon arrival at the base – during the attack my concern was the survival of those around me.

I told them of Turrina's tragic passing, though if there was any effect of emotion on their face, it was hidden well behind static.

"As the most senior station remaining to us," stated Ugden Mirrit, Regent of Commerce, "we hope you have some answers." All business was conducted through the Office of Fair Trade, though rarely would business on Nemi be considered "fair."

His own Mirrit Global had absorbed every investment firm, raising stock prices and gouging the product market. He controlled the banks, regularly raising interest rates on loans – he also held the manufacturing power 0f Khano Industries.

"Where is she?" repeated Hi'Leng Khano, Named Superior of Law, and daughter of Adunaan. "Respond!" She slammed a silent holo-graphic fist.

I was most worried that she was leading this interrogation, holding the power to accuse, sentence, or commit any with no trial. When none other could decide, she would speak the final Rule of Law.

"I could not guess – what would you need of my handler? She too was a victim in this attack, hopefully rescued from SB-3," I implored them. It was The Imprir who had chosen her to oversee my performance in executing their will. Surely they would provide no less than their most invasive vetting techniques to assure her loyalties.

"We believe this conspiracy was perpetrated from within, and that the incident in China influenced by an agent within the American unit. We do not suspect you were involved..." The High Inquisitor raised a hand towards the Superior.

"Correct, though we know that the origin of the shared intelligence was a server within the Dulce facility darkroom," she confirmed.

"The operative assumed to be Jaqueline Stone was trained to oversee your operations in the United States. We do not know where or when they were replaced, only that their true identity is not human," provided the High Inquisitor. This news took me by surprise. I had seen nothing but scrutiny in the eyes of this agent – they had put on a well-practiced performance.

"We do not believe they acted alone, nor that they were the leader of these schemes," added the Regent. "We believe they were recruited through either coercion or by some corruption."

"Corruption?"

"The corruption of their ideals and priorities, by a...*yet* unknown force," came the Inquisitor's answer.

"Were you in any way aware of a conspiracy to inform the Earth populace of our existence?" The Superior was on the edge of her seat.

"Not only this treachery," added the Regent, "but the sharing of specifications for our technology such that they may *rebel*?" This caused a commotion among the other attending members of Imprir.

"Yes, information we have received only recently." A screen appeared above the center of the table, showing what I thought was an empty cell. A shadow crept slowly up to the camera. Suddenly, the shadow disappeared, replaced by a bruised and bloody face.

I recognized the Chinese agent from the reports I read on my voyage to SB-3.

"While the footage was publicized through news outlets on-air and online," the Regent continued, "there were also many key documents lost in the raiding of our Chinese and German offices that were attacked simultaneously. This operative was found attempting to use a...," he looked around him as if to coax the word from some unforeseen companion. "I believe it is called a *fax* of this document, but set fire to any other intelligence they had acquired."

A LightPad was placed before me, revealing the partially destroyed document. The top half of the page was charred, but in the lower left-hand corner I recognized the unmistakable symbol of the Militia Majora. I then studied the text below, reading:

KHANO INDUSTRIES RIFLE SPECIFICATIONS
xx-49729G 22/27

These infiltrators had acquired almost all the pages for a manual to build a rad weapon. It would, by necessity, require schematics for the He-3 power source, an even greater revelation for human technology. If this was evidence of their resourcefulness, they may already have an operating rival to our own weapons.

"Why has this been prepared for their language?" I had many questions, of which this was only the first. The members of Imprir shifted uneasily before the Superior gave an answer.

"You may not remember why we began this endeavor – I suppose it was before your time. When contact was made, the reports were nothing short of horrific. The care taken to mutilate the bodies of our people – dissected as if *they* were the beast – with no respect for the rites of our dead. This was deemed a brutal and uncivilized world. However, our intentions have since changed. Surely, anyone could be rallied to a cause against this great enemy, and the people were so moved. But what was a war of *eradication* –"

"Became an opportunity for our greatest *enterprise*," interjected the Regent. "When we infiltrated the world's many communities, we found they were easily controlled by few of their species. Labor gave them purpose and these desires have allowed us to exploit their conditioning."

These are operations the Regent benefited from, no doubt.

"The planet provides an abundance of natural resources," added the National Dean of Education. If Hi'Ran Moloke was not the most intelligent of our species, his clearance at least allowed him access to more data and sources of knowledge than any other individual. The Moloke family business was that of information and exploration. They also controlled the pharmaceutical companies.

The goal of MoloCorp was to repurpose Khano's technology necessary for sustained off-world missions: air recyclers, water filtration systems, and waste management. They purchased these patents from the auction of Khano Industries portfolio and forced a monopoly on Militia Majora contracts. This left MoloCorp the only company not reliant on Mirrit for shipment of their goods.

It was the only competition left in the Nemitian economy.

"These materials are necessary in all sectors of our industry – but for a few key substances that seem to be unique to our own system, these are materials we have otherwise depleted on Nemi, forced to search for alternative energy sources." Images of what I recognized as the gas giants Saturn, Jupiter, Uranus, and Neptune, flashed across the screen as the Dean waved to his sprite. "Here we can harvest materials we need for vehicles, homes, food processors."

"That is to say that it is not just weapons technology that may be at risk!" cried Hissip Jarrin, the last form projected alongside the Inquisitor. His factories ran the manufacturing industry of Nemi and employed the most workers, though they did not provide for much.

"Currently, as we condition ourselves, we condition them. The parts for this weapon are manufactured separately at factories scattered across different countries we control. They are shipped and assembled only by Militia Majora personnel – none should be aware of their true nature. As far as they know, they work for some car manufacturer or process ore to fuel our technology."

"We have begun construction on the Europa Moon Base and these creatures will provide the workforce we tried to create before the last war, and can now only risk a few –," began the Dean before being muted, the High Inquisitor now leering at the face on the screen. "That will be enough, Imprir Moloke. These details are beyond the bounds of this one's debriefing. We shall discuss this further *after* their departure." Each projected screen faded away.

"I don't understand," I stated simply.

"Orders – *soldier* – are for you to understand. And the good Colonel shall have them in the hangar." The High Inquisitor rose from his chair and walked along the conference table towards his office. "You may have risen through the ranks in the Militias. You may have been chosen to be the presumed leader of some earthly power, but you answer to *Imprir*, Mr. President." He looked back at me before closing the door behind him.

"The Majestic 12 are hereby disbanded," he snarled, closing the door and leaving me alone in the now more gargantuan room.

Upon returning to the hangar, the sound of the grating metal doors drew the heads of the soldiers occupying the room. They quickly looked back to their work, prepping the ship for another journey. The Colonel remained unseen, already onboard.

I looked at the side of the ship, painted with the name *De Quista – The Quest* – in white over a gold crescent. The dominant features, however, were the gold streaks under a large blue quill – the long established symbol of Nemitian merchants and signature of Mirrit Global.

I climbed the ramp and entered the cargo bay. In the single passageway were doors that opened to the crew quarters. Each was a tight fit, allowing one to enter and maneuver onto one of four beds. Four beds and four rooms for sixteen passengers. Just before the control room, there was a door on the left that was a small lavatory.

To the right – now open – was Command quarters.

The only difference between this room and the others was the need for one bed. On the opposite wall, rather than another bunk, was a standing desk. I was glad the Colonel was not waiting for me here – I was feeling claustrophobic.

The door opened to Command deck, and I saw Colonel Stromsky sitting in the command chair with her back to me. She turned, cutting through my facade of confidence. I had seen her leave The Inquisitor's office with the same look. She expected a different outcome from my debriefing.

I had assumed we would be alone, but as I approached the Colonel, I noticed Major Johnson was standing at the weapons panel running a diagnostic. When she turned and saw a familiar face – even if that of a newfound friend – she smiled.

"Kaancha, what are your orders?" she asked, the title reminding me of Nemi. Though the High Inquisitor was the only Nemitian I saw aboard the Moon Base for certain, the language of the consoles and the hum of the CFG would remind any of home.

"*Our* orders, it would seem," deferring to Colonel Stromsky. She sat silently for a few more moments before turning to the Major.

"Did *you* know Jaqueline Stone?" The Major did not and what followed was an account similar to that the High Inquisitor had given me – corporate interests kept us playing in earthly politics.

She also provided the document sample from the office fire.

"Why is this in their language?" the Major asked, repeating my last question to The Imprir.

"Before her assignment to 'White house Chief of Staff,' *our* Jaqueline Stone was to translate the weapons manuals into several languages of which this...English, is only one. Care to guess the other languages?" She handed me a new portfolio. These documents were complete in Chinese, Korean, German, Spanish, Portuguese, and Russian.

Currently, three had been stolen in full.

"This work was legitimate," I growled. "This was the intent of The Imprir, before any espionage was committed?"

"Correct." A smirk twitched at the edge of the Colonel's lips. We both jumped at a loud bellow from behind us. Johnson was now looking at the portfolio.

"*This* is not treason?" the Major cried with a laugh. "*This* is why we're here!" she cried, slapping her leg.

"Careful, Major, you are not the first to feel this way." The Colonel was now glaring at me. "And we know all too well how The Imprir deals with those that question their motives."

The room grew quiet.

"Our orders are to return you to the Dulce facility, Mr. President. You are to find any copies of these documents and any possible avenues of their distribution." She stared out the view port, nothing visible except an electric field occasionally flashing blue across the darkness.

"We will spend tonight enjoying the comforts of Moon Base," she stated of the sterile accommodations and bland rations. "You'll find a gift to share in the cargo bay. Until then, you are dismissed," she ordered with a wave of her hand. She stopped us as we were about to exit the control room.

One last request.

"And I think…*Kaancha*, I think from now on, I like that." No longer was she Colonel Yvonne Stromsky.

She reclaimed her identity – Kaancha of the Mirrit Deployment Vessel *Quest*, Logaani Niih.

Each day more of home was returning to my awareness. I had forgotten the true goal of any soldier – to stop fighting. It was a thought that made me uneasy.

We were only a few steps beyond the door to command when the Major jabbed at my ribs with a gloved fist.

"I guess they're promoting everyone, eh, Imprir?" she joked.

"Or demoting me." While another joke, there is always a little truth in humor – such that the Major's features lost their cheerful expression. I placed a hand on her shoulder.

"Before they may, as President of the United States of America, it is my right to promote *you*, Major Angela Johnson, to Colonel. And Colonel?"

"Yes, Supe?" she responded.

"It's still Mr. *President* to you,"

We both began laughing, entering the cargo bay to find a bottle of spirits.

Historically vessels carried such substances as a means of ritual. It was believed a lost soul should be remembered happily rather than in mourning, though the effects were short-lived. While this was an old ship, this was obviously a personal bottle - the effects would surely last the night. I grabbed two plastic cups from one of the storage containers.

"I'm sorry, I seemed to have forgotten your name?" I realized I had only known the person before me by only half an identity.

"Uh…it is Agamo, Supe. Yekk Agamo," she provided with a confused timbre. "Why?"

"To home, Yekk Agamo," I cheered, opening the bottle to pour us both the first drink.

We exited with the bottle half-empty, shared between us. The other soldiers looked up from their tasks and gladly followed us to the mess hall, eager to commune with their ancestors.

All Militia know a new day brings new risk that they would find themselves cause for celebration.

CHAPTER VII

Dream//MEMORY

The night of reverie had aided my slumber. I rarely dreamed. I could only conjure memories playing out before me. The last was hazily remembered the next morning, as though it were the memory of another. I awoke and shifted my position in the proper bed – if not basic – in my private room on the Moon Base.

Much more comfortable than the bunks aboard the Quest, I thought, as I recalled the Colonel's decision to remain aboard until morning.

The Kaancha, I reminded myself.

I felt I understood the bond between "Captain," and vessel, though I had never had a command of my own. As I drifted off, another memory began spinning its reels.

I found myself back at the desert compound. I could see the basement, but there was another body among those burned away. As I turned to see the face of this corpse, unharmed by the flames, a splendid light appeared and blinded me. Feeling returned to my body as my vision cleared through a static haze. Voices overlapped in my mind and a high-pitched tone whined tremendously, shaking everything around me.

I was now in the position of those cowering on the floor. All around me were again among the living, but I could see that last look of terror acknowledging they were aware of the coming end.

Eyelids drawn back completely, tears coalescing on their edge, and teeth bared in macabre howls.

This violent whistling assaulted my lucid ears. I awoke hitting the floor, wrapped in my bedding. I had knocked over the small bedside table and the clock sitting atop it. The alarm had been blaring, but the whine was cut short as the casing shattered on the hard floor – Militia grade plastics, not made to last.

I organized my bed sheets and donned my uniform. I had not worn proper garments befitting my unearthly rank since my transformation. I found the fitted jumpsuit comfortable and as devoid of medals as a fresh recruit. With each passing day, I was less in control and receiving more orders than I was giving.

This brought both relief and worry.

I had been left to my own devices with minor oversight until The Imprir decided it necessary to install Jaqueline Stone to the White House – a decision they now regret. The necessity of my presence off-camera and outside of the Oval Office dwindled as my new Chief of Staff was adequate representation. I was in the middle of my elected term, but I felt paranoid that I was no longer needed for their grand scheme.

However, this could have been a side effect of the conditioning.

As I entered the hallway, I found the other rooms remained latched – the sounds of their alarms could be heard unfettered. I headed for the hangar, hoping to have a few minutes alone with the Kaancha.

The loading bay was completely empty. There were no fuel regulators, no charging stations, or any of the many instruments deployed by a pit crew when outfitting a vessel for its next mission. The only items visible were the containers unloaded to improve fuel efficiency by decreasing the total weight of the vessel.

An alternate reasoning would lead one to believe the extra space was being made for more trained crew. While I did not yet know what questions to ask the Kaancha, it was this alternative that gave me direction in the upcoming discussion. I prepared to have the first and last word before releasing a latch that activated the loading ramp.

The ship was simply *off*.

It was completely motionless, not a single vibration from any running engine. There were no sounds – not of fluids pumping throughout the hull, nor the buzz of electricity. In space travel, there was a need to recycle air, the components constantly at work and emitting a sharp hiss. If these recyclers were not functioning, the oxygen content of the air supply would become depleted, replaced with a fatal ratio of carbon dioxide.

The Kaancha may no longer be giving *any* orders.

I rushed through the tight space that was the passage connecting this bay to the control deck. I stopped to open the door to Command quarters and found the bed empty and unused. This moment passed without my missing a step and as I entered the cabin, I could taste the stale air though I could not determine how long the recyclers had been off.

I knew I needed the Kaancha to survive – if only to answer my queries.

Slumped over in the Command chair, I observed the still form I knew to be Logaani Niih, her prize bomber jacket draped over her.

I approached with a grim expression.

How could this happen? Was this evidence of more sabotage that had followed us off-planet? Or had the eccentric Kaancha finally reached a breaking point in the night? I saw another bottle of spirits beside her, empty. As I stood behind the seat of command, I reached over the chair and removed the jacket. I placed a hand on her shoulder and squeezed it to pull her back and confirm my assumptions.

Before I could do so, however, Logaani jerked to life, clasping her neck with one hand and my wrist with another. She looked around in a confusion and I believed she was suffering from asphyxiation. I pulled her to the ground and began applying CPR techniques.

I opened my eyes after exhaling and saw Logaani's were wide open. A powerful fist shook me from my position beside the prostrate Kaancha and it was I who now found myself on my back.

The Kaancha leapt to her feet, one hand still holding tight to the left side of her neck. The other slammed the panel before her and activated the primary systems. She then donned the bomber jacket in a swift and practiced motion.

She flipped the collar up and put her hands in the pockets before turning to me.

"Who is in command here?!" I cried from my submissive position.

"The Imprir is in command – they are *always* in command." She corrected the presentation of her uniform and the pin denoting a Kaancha – a gold crescent with three points bored through it.

"To whom belongs the title of Superior?"

The door opened as the other soldiers had boarded, Yekk in the lead. They coughed in the toxic air and brought their hands to their mouths until the recyclers performed their function.

What they saw was a powerful leader garbed in what was considered spoils of war – the light from the panels glinting off of a gold pin. On the floor was a less impressive figure in the same uniform they wore, with no defining feature of rank.

"Perfect timing," I heard the Kaancha mutter to herself before proclaiming for all to hear, "You can call me, 'Supe,' then, I think," and loosely saluting before helping me to my feet. Kaancha Niih turned to the crew and addressed them as though I were not there.

The distraction provided me with a moment to check the toxicity levels of the air at the moment the Kaancha reactivated the recycling components and other major systems. What I found seemed impossible. I looked back at the Kaancha, believing it must be a glitch. While the atmosphere of Nemi is more robust than that of Earth, the stale and bitter air should have been unbreathable.

I stepped away from the component just before the Kaancha turned back to take her position at the center of command.

"I had more to discuss with you, *Supe*." She glared at me before relaxing and smirking.

"*Mr. President,* I shall meet you in my quarters after we take our leave of this base." The other soldiers were now manning their appointed stations. Yekk exited to wait in crew quarters or the cargo bay.

* * *

"*You?*" I cried at the continuously smirking face in front of me – an expression that was *not* growing on me.

"Imprir orders. I'm leading the away team as well. Omega Purseis can handle the *Quest*."

"But we are still headed to the Dulce facility?" I tried to glean more information on our mission. I was now the subordinate, the burden of mission details no longer mine to bear.

"Correct. We have retained the five codes provided by those that made it to Sub Base-3. We have also confirmed that the remaining seven have been changed at their sources. We no longer have control."

"Why aren't we trying to get the two codes necessary to regain control over the planet's weapons? Have we not prevented the manufacturing specifications from being delivered?"

"Correct again, *Kaancha*." I took note that, although she would not use it in front of the crew, she spoke our equal rank with derision.

"However," she continued, "The Imprir thinks the impostor had help. They believe the original documents she was provided – by the Majora, let us not forget – remain in the dark room."

I suddenly recalled that the Jaqueline Stone overseeing my operations in the White House had introduced herself to my companion world leaders in that very location. It was apparent there was more to this room than the view screens and communications panel – she had remained in the dark room alone for several hours.

She never entered the room after that, not that I was aware of. This all but confirmed she was replaced early in her tenure at the White House.

"She may return to complete her objectives."

"Once we again have control of their nuclear armaments – should we enact this contingency – some will survive. We will leave them no hope. We don't want any resistance in the last days of *Loarkeen*," she said, using the newly written Nemitian word for *human*.

"They must not retrieve these documents," she asserted with the finality I had become accustomed to as a Superior.

"I understand. I must return to keep up the farce, but why must I assist you in *your* mission?"

"The impostor could not access the dark room because she was lacking what the original – and yourself – should not be – your personal ID chips."

"The Imprir could have given you clearance." I wanted a return to my regular duties.

"The Imprir would like *you* to complete this mission and prove your worth." She waved me away and turned back to the reports on the small desk.

"You are dismissed," she ordered as I exited her quarters.

I made my way to the back of the craft, returning to the cargo bay. I saw Yekk Agamo and several soldiers – just enough to fill the bunks – gathered around a communications panel. They turned and glanced away from me as Yekk ended the communique. All but my new comrade anxiously exited the cargo area.

"More specifics to our mission?"

"Not ours, Supe. The Kaancha and I will be deposited on the compound to aid you in recovering the weapons manuals. Omega Purseis will prepare the rest of the crew in the event of a hostile exit."

"Are we expecting trouble?"

"We don't know what to expect, Supe."

"Are you aware of any reason to remove the cargo that once filled this hold?" I questioned her, motioning to the empty space around me.

"All I can say, is that we will rendezvous with a second team after completing our objective." I could see that she was instructed against sharing this information with me.

With no direction for this conversation, we stood silently and acknowledged our mutual displeasure with the situation. After a few tense moments, the communications panel was activated from the Command chair, projecting a small image of the Kaancha's features. The ship hummed and vibrate with the activation of the CFG.

"This is your Superior, prepare for take-off," she ordered. "Our mission starts on the other side of the hangar gate."

Yekk and I took a seat on the bench I had used to rest the previous trip. We strapped ourselves to the wall and Yekk placed her chosen weapon in a wall compartment above us. We sat uncomfortably, though unrelated to the shaking of the ship.

I knew Yekk was a soldier, just following orders, as truly as I knew she questioned those orders.

CHAPTER VIII

TR3-BETA

The return trip was not a silent one.

Once we reached open space, we were free to move about the ship. Yekk and I joined the other crew members in the control room and I was tasked with mapping the compound – intel I was reluctant to provide. If the true purpose of this mission was to recover documents from the dark room, why was a full structural blueprint required?

Why were we scheduled to rendezvous with a second team?

I had assumed a successful mission would place me in my previous position as President of the United States, ever working towards control over the industrial and military sectors. The Kaancha and her team would have no more reason to stay than I would to leave.

This may truly be the end.

I was a soldier – my character was the politician. I had always seen this endeavor as another mission. There would be no war after the Imprir revealed themselves to be the gods of this new world.

Their prize would either be destroyed, a wasteland lacking of life to be stripped to its core, or its peoples enslaved, forced to dismantle their world to the benefit of another. Once the Earth provided no more, they would be sent out among these new stars to serve as slaves for eternal generations.

I could not decide for them which fate was the greater for suffering – to exist to serve, or not to exist at all?

I returned to the cargo bay alone. The Kaancha announced we would first establish a low orbit, though I knew not why, and I resigned myself to a new fate come tomorrow as I strapped myself in.

I had led Militia crews in every capacity – on as many missions of exploration and science as combat. There had been no need for conflict in my enlistment career until the few encounters with native forces necessary to keep our secrecy. In my youth, I yearned for new adventures like any other youngling.

I had now been living in the United States for thirty-three years. I had come to Earth in what was 1976 with a group of recruits from the Old Militia. During Instruction we were divided by rank, and I was grouped with the other volunteering Kaancha. This included myself, Turrina Tua, and Chirrak Pakris – both lost during the Sub Base-3 attack. Also present were our comrades Iniki Atavan and Uubiron Shikase – Rajesh and Adebowale – whose current location remained unknown to me.

The last member of our group was not a Kaancha, but had shown much promise in tactics and soldiery. This was Logaani Niih. She was deemed unfit for a leadership role due to some detail in a lost psychological evaluation.

Others in our group that could not complete the rigorous mental conditioning required to be placed in such lead infiltration positions were moved to another group. They would be trained for movement, with less personal relationships required to maintain their cover. It was this team the original Jaqueline Stone would have been assigned to.

Volunteers of lesser rank and civilian recruits, such as Misters Walton and Castello, were trained in a separate conditioning facility. While the grounds of our base contained several structures, the civilian simulation resembled a small town. Native peoples had been collected to provide a control group to observe. Each night, they were reconditioned to repeat their scripted tasks. This provided a metric to judge how convincing our performance was.

While I had not started a family as part of my cover, I had grown accustomed to having a home. Many who surrounded me were natives who existed on the periphery of my awareness, our time together a function of my station – some had left lasting impressions.

There was a humor and wit to humanity. Beyond the ability to adapt simply to *exist* is to carry on through the void. When a species evolves beyond the need for this survival – truly to be – the mind can dwell on the importance of purpose and what that purpose might be and find an answer.

Many sought this state through meditation and other ritual practices on Nemi.

Those who have succeeded expect that the same comes easily to others, while for many, the search for purpose is a lifelong journey. I have seen an optimism in the peoples of this world to find a unified goal – the greatest threat to the Imprir.

We could infiltrate the ruling powers because the political realm was already a dark stage, ripe with deceit. It was my belief that if the people seize control of their lives by taking ownership of the industry they labor to produce, this would be a prosperous world soon to rival Nemi.

This is what we were now dispatched to prevent – our mission to intercept the means of this revolution.

Did I want my legacy to end alongside humanity?

Would I be remembered as a soldier, just following orders? Or were there moments when authority was meant to be disobeyed? Was it the responsibility of history to identify these moments, or is every individual responsible?

These are questions soldiers should ask after the mission is complete, having faith that the *cause* – belonging to no one individual – is justified. It was this faith I was losing in the motives of the Imprir. I no longer believed this cause to be the will of the Militia Majora, but that of the corporations.

The only communication with anyone off-world was directly with the Imprir – did Nemi know what was happening on this side of the Break?

Stories were told of how the Kaancha and crew of the *Dark Eye* discovered demons waiting to corrupt our world in a realm beyond – a realm with so many stars that could be home to more potential enemies. They returned from this now mythological journey almost a century ago in modern time.

I was still young when I enlisted to slay these demons for their perceived sacrilege.

The religious zealotry and fear that led the Old Government were removed when the first Imprir appointed Hi'Ran Ahket – then Speaker of the Temple – to High Inquisitor. What began as a divine expedition to purify an evil realm became an opportunity to find a new slave caste.

My personal fear of this world had diminished with time. There was much technology Nemitian scientists could assimilate into our own. While we had telescopic instruments, until the appearance of the Break, the only stars and worlds within our awareness were easily visible.

On this side of the doorway between worlds, to view the wonders of this universe required scrutiny we did not possess. While I had studied the native research into extraterrestrial life, I viewed much of what can be seen from the impossible pearl of life that is Earth.

I had once believed we should leave the planet behind and focus our efforts on exploring this vast cosmos for greater treasures and allies. Earth's leaders had long hidden the truth from the population. Their industry had made no preparation for a defense, nor had they been motivated to drive forward the technologies necessary to travel these many worlds themselves.

As we settled into orbit, I heard the door beside me open with a *WHOOSH!*

I saw Yekk's head extend beyond the boundary, looking both ways, and I smiled at her. She remained expressionless in return, releasing the clamp keeping the wall compartment containing her personal weapon closed. She removed the weapon, strapping it to her back with the green cloth. She stepped aside and stood at attention on the far wall.

Whatever sudden issue she had with me, I attempted to keep a friendly face.

Soon, the cargo bay was filled with the other soldiers, last of which were the Kaancha and the second in command, Phalae Omega Purseis, close behind. His features gave the impression he shared the Superior's opinion of me. The Kaancha, smirking, turned her back to me to address those gathered along the wall. I couldn't hear the Kaancha's instructions and I was getting irritated with the decision of the Imprir to keep me uninformed.

If I wanted to learn *any*thing, the initiative was mine to take.

The moment my patience ended, and I walked across the hold, the Kaancha raised her gloved hand. The air directly in front of me shimmered bright gold, taking on the form of a new vessel. I leaped backwards in surprise and tripped, falling onto the hard metal floor.

This scene only reinforced the loyalties of the crew.

I resigned myself to perform nothing more than my duty on this mission. I joined the others on the opposite site of this new craft, the results of our efforts to retrofit human and Nemitian technologies. It was dubbed the TR-3BETA model – no titles for this stealth craft, forgotten if lost.

The plan was as follows:

We were to land both cloaked and under cover of night near Dulce, New Mexico. The Kaancha, Yekk Agamo, and I – the presumed President – would depart. Our goal was to collect the documents from the dark room at the facility hidden under the desert.

"Before collecting you at Sub Base-3," the Kaancha informed me, "I had to dispatch your personal secret service members, trusted appointees of the Imprir that were replaced by impostors. We have reason to believe they may have infiltrated the compound's guards."

The TR-3BETA would remain cloaked and the other soldiers had orders to maintain a perimeter. Only Jaqueline Stone and I were aware of the dark room. It was unlikely it had been discovered, but the Imprir were prepared for resistance. Omega Purseis would remain on the *Quest* alone awaiting our return, all other soldiers aboard ready to be deployed.

Appearing near-human, this was the next class of NIB bound for Instruction.

We descended below the clouds before activating the cloaking device – it required an external battery and one would not waste such an advantage. The fore of the ship, visible through the view port, shimmered across the spectrum before the cloak caused it to disappear from our vision. It was replaced by a digital representation projected onto the glass, along with sensor information.

As we streaked across the landscape, passing over fields of grain in a blur, they truly appeared as amber waves.

We crossed Illinois, Missouri, and Kansas. Looking back to the West, the Sun's last rays curving through the atmosphere, the mountains truly glowed a majestic purple hue.

We entered New Mexico airspace with no trouble and slowed our descent as we passed over the mountains, invisible to any in the area as the craft narrowly avoided the tree tops.

What a strange breeze, they might think to themselves, ever unaware of the enemy moving silently on the wind.

Approaching the local Air Force base, we witnessed several vehicles and bodies hunched over telescopes – the abandoned base became a popular UFO hot-spot during its use. We were low enough to cause a few hats to blow over the sturdy fence bounding this government claim to keep space from the public.

The TR-3BETA rested lightly on the dirt and the doorway opened. Passersby would have witnessed a light from nowhere, much like the Break that connected our two realms.

"After you, *sir*," ordered the Kaancha, motioning that I lead, followed by Yekk, then herself. The NIB filed out and marched off in several directions. The three of us stood by as the door closed, the ship completely invisible.

I stood in my jumpsuit, while Yekk and our Superior were in standard stealth ops fatigues - all black and of a material that was invisible to infrared. The bomber jacket I associated with the Kaancha's presence had been left aboard, as if an old skin being shed.

CHAPTER IX

Choice

It was classically too quiet.

I had an intuitive feeling that the NIB ordered to establish the perimeter were tasked with eliminating or corralling anyone on the premises away from the main hangar. The crew was setting up for an ambush.

We continued on to the single concrete structure housing a lone elevator. We descended until the rock walls fell away, exposing a large natural cavern. Within a small town had been constructed, soil and trees transplanted to form a forest around it.

In the center was a church, across the street city hall – our target.

The Kaancha's face was ever smirking. Yekk's features were contorted, attempting to force a facade of focus. This made me hopeful she was as uncertain about her role as I was. I could hear crickets chirping and the wind generators rustling the leaves of the trees. This would be our cover as we remained crouched and light-footed, though I doubted it was necessary.

The Kaancha kept me in front of her and in the lead position. Yekk and I waited for her order to vault over the bushes. Yekk gripped her personal rail gun, the Kaancha a pistol.

It was only now I realized I had not been provided a weapon.

The Kaancha's face was ever smirking as she read my expression, my emotions exposed. She signaled us over the railing.

We planted our feet softly. Lights were shining from every window, but I heard no voices or echoed footsteps from the halls. I opened the door to the office, leading the other two along the wall until I felt it fall away, my hand warmed by the heat of a false barrier.

I pulled my ID chip from the breast pocket on my jumpsuit and exposed the wall paneling that would confirm my identity. It scanned this chip and a DNA sample provided by a shallow, painless prick of the thumb as I placed it on the panel. Usually the ID chip would be enough, but due to recent events, the Imprir locked the dark room completely.

The usually blue back light changed to red, and a waveform appeared oscillating across the display as more data scrolled past a photo and my name – Oscar Orwell.

I had lived a life of titles, of which. 'Mr. President,' is only the most recent. I rose to this position honestly, serving as Congressman in my last performance under the Majestic 12. In the Militia, like any military, names are lost, replaced by rank and service codes. The only title that truly mattered was, 'Superior.'

"Oscar Orwell…," I breathed unconsciously.

This name had followed me since the day I first entered the facility below the desert of New Mexico. Those of us chosen to infiltrate leadership positions had to start at the bottom. Because of a lacking responsibility taken to assist the homeless – adults and children alike – it was a simple task to falsify documents establishing citizenship and birthright.

We forged many identities.

We were genetically altered to appear as our enemies before taking Instruction on planet. There were no permanent bases other than that outside of Dulce, and it was during this early period of our infiltration we had the highest risk of exposure. We were fully reliant on native contacts that were later disposed of. The risks inevitably realized themselves. NIB were deployed to solve these problems locally and abroad. I took the name Oscar Orwell after my last mission in this service.

The next step in my infiltration was attending American high school, where students are tested and future opportunities are provided to those that test well or are financially capable. Naturally, I excelled at the primitive science and technologies courses. This gave me ample time to focus on American history and international studies.

It was a simple path for a dedicated soldier to follow. Those I surrounded myself with thought themselves good friends, though it was an act. It wasn't until I attended the universities of higher learning I allowed myself – to allow Oscar Orwell – a friendship. I was pledged to a fraternity, knowing the connections I made among those privileged enough to share in this brotherhood were the most sure way of entering the political realm.

As I had served in the Militias, I was chosen to enlist in the Earthly militaries, shipping out overseas. Uubiron and Iniki – now Adebowale and Sergei – also ex-Militia, were chosen as well. Iniki and I performed joint operations with Uubiron in Nigeria. I now remembered having met the Major – Yekk – and making note of her secret weapon. Though the rail gun was too advanced to deploy in these skirmishes between natives, she always kept it nearby.

We served honorably, even if it was a farce, and I became close with many of the soldiers in my unit. These connections were real but would also serve me in future political campaigns, as well as those connections made by Uubiron and Iniki.

"Your Presidential hopeful – war hero and friend of the working man – Oscar Orwell – Double O!" I remembered of the introduction I had been given on the campaign trail. My advisers said using this college nickname would make me more relatable to voters as we traveled through the South. It was nearly November, and they were looking to bring us up four points, leaving me behind only six. The next day, we were *up* by four.

I had been standing with my hand motionless above the panel. A menu box had appeared asking that I confirm disengaging the barrier and revealing the dark room beyond. The Kaancha was looking around nervously. She was tapping her foot rapidly, betraying her extreme impatience.

"What's wrong? Does the ID chip not work? Did you forget it? What? What" Her brow was sweating and I could not understand why she was suddenly so stressed.

I saw no change in her condition as I pressed the panel, disengaging the false wall, a doorway opening into the communications hub. Yekk surveyed the dark room – weapon at the ready – but it had remained hidden from human eyes. The Kaancha rushed past me and I noticed a scar on the back of her neck. It was behind the right ear and jagged, leftover by a botched body modification.

It was a scar I had become familiar with.

"Jaqueline?" The Kaancha turned to me as she realized my discovery. She shot me with a quick draw of her pistol.

I knew it must be set to stun as I fell. I heard a second shot and the grunting of another soldier betrayed. Yekk's weapon clattered to the floor next to her.

Even at the lowest stun setting, I should have lost consciousness. I remained disoriented, however, as it was difficult to lift my head to observe my surroundings. A red light I was unaware of illuminated the room, allowing me to see Yekk was unconscious, though I knew not why I had been affected less. I watched as a shadowed figure pulled several of the wall panels I had previously assumed were permanent fixtures off the wall and throw them away in a mania.

I saw the shadow pass from each of the exposed compartments, throwing out documents that were now scattered across the floor of the dark room. Whoever this imposter was, they had not expected me to regain my bearings so quickly.

I was still, not wanting to lose this advantage. Moment after moment passed as this unknown foe rummaged through the mess they had created, searching for the rad weapon schematics.

I could sense the tingling of feeling and control returning to my body.

The imposter searched nearby, but I remained motionless. This provided an opportunity as they turned away from me and growled in frustration. I grabbed their ankle and leapt up, righting myself and simultaneously downing this enemy. From this power position, I braced them against the floor and began my questioning.

"Who are you? Where are Jaqueline Stone and Logaani Niih?" They spoke quietly such that I could not hear them. I leaned in and asked, "Where *are* they?" This time, I did not miss a word.

"We're right here," came their reply, followed by a hard strike from their helmet adorned head. This left me disoriented, allowing the imposter – regardless of who they claimed to be – to escape my hold. They searched for the blaster which had been ejected from its holster when I flipped them over.

It had slid far enough across the floor that it was within Yekk's reach who had awoken during the scuffle. She crawled towards it, but was too slow readying the weapon to keep our enemy from snatching it out of her hands. They adjusted the setting, raising it to inflict pain but not to be fatal. Most would deem this cruel and unusual, a setting rarely used when the options to incapacitate or kill are available.

They blasted Yekk, who let out a cry before dropping motionless to the ground. The enemy turned to me, weapon raised, and walked across the room.

"Today's the day – big day, big day!" Their hand reached into a pocket and retrieved a conditioning device. They dropped their pistol to free their right hand and grasp my hair, lifting my head. The left hand shaking, they raised the device to my eye.

"Time to remember, time to wake up! Or time to dream again...," I heard as her voice faded away.

My eye was assaulted by a blinding series of flashing lights. I had recalled no conditioning of this level, but I suppose that was the nature of the procedure.

This was not a wrist device – it was an early model resembling a small bat. Before the technology had become advanced enough to encode an entirely new identity over an individual ego, it was used in therapy to recover repressed memories. In special cases, it has been used to repress or alter those memories too painful to remember.

It was one of these altered memories – a dark parody of what I already considered a monumental failure and a moment of weakness – that now played out in my mind.

* * *

I found myself in an infinite space. Many voices drummed in my head as deeply as the strobe patterns that still burned my eyes. I was lucid but unconscious. I was aware of who I was – a decided improvement from my previous visit. There were no defining features in any direction, but the emptiness was familiar.

While I was confused why I continued being pulled back here, I took comfort in *this* unknown. There did not seem to be anything to know but the drumming of fluids in my skull. I was beginning to understand this reality more than that I was living in.

Here, it was enough to simply exist.

Again, it felt as though no time was passing. I knew I was within my own mind – I knew my imagination was limited, bound by my experience – but I dreamed big. I dreamed beyond the Break since I was a youngling, drawing what I interpreted alien worlds and their inhabitants might look like. Endless nights, I lie awake and stare out the window into the cosmos of humanity, no less overwhelmed than the night before.

When all that can be seen is all that is believed to be, one can easily become content with their station, but the ethereal void encompassing the Nemesis system could never compare to this canvas littered with stars, a vast ocean foaming with life. The monotonous mechanical nature of Nemi – one moon orbiting a single gas giant in what we now knew to be a common binary system – will be difficult to return to.

While our technology was superior, I often wondered if my life was improved in the here and now. I yearned for what humanity affectionately calls, "simpler times." But a doorway to endless new worlds has been opened before me and I could imagine every single one of them. As my own life passed, if I took a side step I would find myself in a new theater and watching a different film.

When it ends, I expected I would step into an entirely new life.

I noticed another presence in this space, more interested in observation than watching the most important moments of my existence projected before us. I could feel the intent of this being to communicate like an annoying itch in my ear, but their attempts were incomprehensible to me.

I called out and was met with no echo. I attempted to greet the presence, though I felt absolute nothing – the great loneliness of this place. The throbbing of my physical form subsided.

The hair on the back of my neck raised as a hum escalated in my ears. The presence had returned with a hunger to speak to me. So terrible was this appetite that their attempts caused a great pain in my head. Something was keeping us from communicating in the way the presence desired, and they retreated. Slowly, the pain subsided and I recognized a voice not my own.

In that moment of recognition, a face manifested itself in my vision. It became more real as dark hair grew from its head and a body took shape below it – this remained obscured, as if out of focus. The features now defining the face, however, were unmistakably those of Jaqueline Stone. Blue light was being emitted from the form, looking upwards as it streamed from the eyes and mouth.

"Who are you?" I asked, surprised to hear my voice again.

I assumed this was some manifestation of my psyche attempting to assist me in overcoming whatever weakness my conditioning had afflicted me with. The figure spread its arms and looked upon itself, unsure how to answer the question.

"This is one role I've played – of many – but this name you have given, *Jaqueline Stone*, is incorrect." They looked around as if aware of another presence.

"*What* are you...," I asked, more to myself than the being before me.

"You...You are familiar with this one. You find this image is more palatable? But not by this name." The light seared through me as our eyes met.

"Then what shall I call you?"

"The name I hear – Logaani Niih, *Jackie* and I are one and the same."

"You're hurting me," I pressed the entity. The pain immediately ceased.

"You are waking up, and I am here to help you acclimate to a new reality. You have to make the choice – only you can choose to walk a path of your own making."

"I don't understand," I replied – a claim all too common as of late.

"You are being given a chance to fight for this choice."Another riddle.

"What *choice*?" I desperately asked.

"The choice to live the life made for you, or risk the end to be free – to live one that is your own!"

"How can I make such a choice?" I grew impatient with these vague statements.

"You have once before – now you must choose to remember true."

Two wars exploded around me, one of a new life similar to the one I knew.

The scenes before me became clear and directed, pausing at one particular memory. I saw two realities – my vision separated by these two worlds. On the left was a story I had seen before, that I had been conditioned to believe. It felt safe, yet a fear within urged me to look away. On the right was a hazy projection of another story I had been told – that of betrayal.

I made a choice – I dove through the surface layer.

Underneath this false vision, the truth I had so long ago forgotten was now revealed.

CHAPTER X

Vietnam, 1970

.

.

..

....

RUN//MEMORY.Override
RUN//scenario/08171988
RUN//Engage.Override

Pakistan, 1988

Uubiron and I were strapped into the chopper. We had received our target coordinates and taken to the air fifteen minutes before from Islamabad. As we flew over the desert – close enough to disturb the dunes below – we noted several blasted camps and continued towards the sunset.

The drone strikes were a success.

We would land at dusk, our mission completed before dawn. I looked at the surrounding soldiers. We all wore black suits and hats to hide our altered forms. I did not recognize them and when I tried to identify the pilot, my vision grew hazy.

Their forms shimmered for a moment before I found myself alone in the chopper.

* * *

Vietnam, 1970

Uubiron and I stood on the bank of the Núng River watching the sunset. The sky was a deep purple gradient to pink on the horizon. We had gathered many to support our cause. When faced with global extinction, the natives were more than willing to arm themselves, but a military force this was not. I was glad to have a friend and fellow Kaancha at my side, along with some of our fellow NIB. It was relieving to find others that shared my view.

This vast jungle provided the perfect location to recruit from within the ranks of the American forces already working with us to gain a foothold from which to infiltrate China. The soldiers were ordered to commit horrible atrocities, eradicating villages in a blaze of napalm. The guilt drove them against their superiors and the military at large. The locals were afraid of this invasion, and both were led to join our cause – to free humanity of the shadows that would see it destroy itself.

After revealing our identities, we were seen as gods. Our followers were easily controlled, but any fanatic eventually becomes *fanatical*. They would do anything to please their perceived masters, though admittedly we boasted the superiority of our species. I wondered if we were abusing these people for our own interests – no different from the Imprir we wished to rise above.

"No, Kriss, we are not like them. To the Imprir, *we* are superior," Uubiron assured me. "We may lead them now, but we will lead them to their freedom."

"Like any good war, we start in the trenches." I slapped my comrade jovially on the back. "It's not their fight, really, is it? We've been un-knowingly losing our freedoms for generations on Nemi. We've allowed our leaders to be held hostage and sell us to the highest bidders."

"For what?" he asked the stars above, growing agitated. "What di-rection are *our* people going? We now have this endless space to explore.

There are surely entire planets of E-79 we could harvest resources from, but we engage these primitive peoples – to be thrown away once we've stripped their planet of resources as we did our own?" He paused, looking off into the surrounding trees, a nervous look flashing across his features. He shook his head, put a hand on my shoulder, and headed towards camp for communion.

We then returned to our base of operations, several large tent structures that were our camouflage and defense from the elements. Most of our "soldiers" were malnourished and mistreated youths – refugees. We provided for them all, hoping to protect them from the war in their homeland.

I opened the doors to our sanctuary and was bathed in flashing multi-colored light. I turned away, but the pain wriggled across my eyelids. Even through the blinding flashes, I could see I was somewhere *else*.

* * *

We approached the target location and searched for any survivors. The Colonel ordered they be captured alive before she entered the compound, fortified concrete barriers composing the walls. Several insurrectionists carrying equal firepower to our own had patrolled the grounds, now buried under the rubble.

Uubiron had fought for the party to equip rad weaponry but the Named Superior had denied the request, claiming it presented too great a risk of exposure. Uubiron felt allowing the possibility of escape would be the greater fault. A list of operative code names had been stolen with the intention to distribute them to several locations – all intercepted just hours before. These identities had been culled for years while we underwent several procedures to appear and function as these humans.

As we searched the compound, we crept among the blasted ruins. I saw the grains of sand from our footsteps rise and travel the winds across the burnt and bloody faces of the dead.

As they were swept into the air, however, they formed a great whirlwind that encompassed the camp. I heard the voices of my comrades calling out to me as I sank beneath the sand at my feet, now liquefied. The ripples raced away from my turmoil. When encountering the structures of the desert camp, the disturbances consumed them as well. The walls shook themselves apart as my lungs filled with sand. My form was crushed and compacted, but I would not die – only pain.

* * *

Our followers gathered in the main structure, an open area covered with many camouflage tarps. Scattered about were rudimentary chairs and tables, now occupied by over twenty bodies – not including those offspring held in the arms of their mothers.

The camp now had four children among us.

Each night, at no particular time, we held a communion. It began by accident, those pledging their loyalty to us awaiting some thought-provoking speech each time we gathered. This continued expectation was proof enough we had not convinced them fully of our mortality.

"We are not your superior." I wished to assure them that their best interests were our own. "We come from another realm, and it is true our technology is more advanced than your own. But we are *not* gods."

"We fight amongst ourselves even now on our home world. Some suffer from poverty, famine, and disease. We think ourselves free nations, but invisible hands direct our every action. We have lost our *choice*, and those that took from us – from their own people – now wish to take that freedom from you!" I cheered, but the crowd before us showed no excitement. They simply stared at us, mouths ajar, their eyes begging for more.

I looked over to Uubiron. When he turned back to me, eyes wide and charging, I saw the shadows behind him in the marsh.

"Kriss! They've found us!" Uubiron cried as I heard the first shots and last screams echo endlessly as I was pulled from this memory.

* * *

We saw no one within the bounds of the outer wall. The Superior motioned for us to advance on the central structure, a large concrete home with barred windows and a tarped roof. There were several huts organized along the wall on the inside – all destroyed in the aerial strike.

Admittedly, the group assembled here were not highly trained militants. They believed themselves revolutionaries until they learned the truth of the many worlds out there, of how small this world truly was. The traitor allowed them to believe our kind were gods, such that they would remain obedient. The humans are the enemy, but the traitor's desire for such devotion was a crime contrary to our base moralities. Whether the means are religious or industrial, we believed no one individual should hold so much power over another.

Our squadron surrounded the building, Uubiron and I following the Superior to the front door. A few steps more and we would begin our raid.

"We've got one!" I heard someone call from the roof.

We found the door braced. The Superior set a charge, and we waited for the blast around the corner of the building. The squadron collected behind us. They had not found a back door – the enemy had no escape. The small explosive shook the building, knocking rubble out of the walls and off of the roof onto our heads.

We charged through the swirling dust cloud into the structure. The enemy fired first, and we replied in kind. One of our party was lost, but we had them severely outgunned. Our two armed opponents went down with ease. As I looked upon their pained bodies, I regretted not doing more to requisition the rad rifles.

One soldier raised a bloodied hand, as if reaching out to *anyone* for help.

A shot came from behind me as Uubiron ended this misery. We then heard footsteps and the shedding of tears from above. Our target had not been identified among the bodies and I saw the Superior lead two more soldiers up the stairs.

They ran through the solid wall as it turned to sand. The floor rose and swirled around me, slicing away until it was cutting at my skin.

Fight the pain, a voice called out in the gale.

Remember true.

* * *

They had surrounded us before Uubiron and I could warn the others. Only two of our number – not including Uubiron and myself – were Nemitian and served as defenders of our encampment. They were the first to fall in this ambush, their translucent blood shimmering in the patches of sunlight illuminated through the trees.

The camp was fortified on one side by a wooden wall. The members of our party that remained now rushed in this direction, some remaining to help the children to safety. On the other side of this barrier was a buried concrete bunker. We had hoped it would serve as a storm shelter and only when necessary.

A blast rang out through the forest after several of the refugees had made it over the wall. Those atop the structure and attempting to climb it were obliterated or buried in the explosion. Some had been thrown from the area, the blast having taken more than one limb. One was conscious, screaming in pain at the bloodied stump where a hand should be.

I was surprised to hear mercy cut through the wailing, ending the suffering of its source.

Through the swirling cloud of smoke, I saw a figure in Militia fatigues approaching. I could not identify them as they wore a helmet, tinted visor down. They were holding a weapon as yet unused in this ambush – a flamethrower. I raised myself up as a wildfire erupted around us, dashing flames leaping between my attacker and myself.

Just before the flames obscured them completely, the figure removed their helmet, revealing a familiar face with a wild scowl. Our pursuers were being led by Logaani Niih, her special ops group famously unknown to all but Uubiron, myself, and those who shared our station in Militia command. She cried out in a rage as I weaved my way through the walls of heat.

A pain in my neck caused me to collapse as the flames reached out to me.

* * *

We ascended the stairs to find it empty. A window was open and ladder extended to the roof where the lookout had first announced our position. Uubiron climbed the ladder first, looking back at me and shaking his head. They had escaped.

"A zipline. A *zip*line, really?" Our Superior joined us on the roof. They had removed their helmet, and I recognized Logaani Niih from Instruction.

"What is that?" Uubiron pointed down the path of the wire. A concrete structure was buried in the sand with the door exposed.

It was the only way in or out of the bunker.

"How f– f– fun," stuttered Logaani, raising a special weapon we had brought to use in the cover up – a flamethrower. She squeezed the trigger and a powerful flame erupted into the air. The fire burned a bright green, enhanced with our own fuels.

I heard a loud hiss and my comrades disappeared in the flames, growing to encompass all around me.

* * *

They had chased us from the camp on the riverside and into the jungle. I looked back to see them burning everything – a tactic I knew should cover up our rebellion. If they had found us here, then they had surely found the other camps. We had been betrayed, though it was possible the traitor had become a casualty in the blast.

I searched for Uubiron but could not find him.

I hoped my comrade had escaped over the wall before the explosion and I would find him safe with the other survivors in the bunker. The wall had collapsed, partially barricading the door to the safe space. I did not know if anyone was inside, but I could not let them be trapped.

I charged the larger limb that had fallen over the door, bruising and burning my shoulder. It was not severe enough to slow my pace, though I howled as I grabbed the handle of the metal door, burning my palms as I threw the door open.

I entered the bunker, and I found Uubiron with several refugees and three of their children. There was not much time before the hunting party would be upon us.

"We have to go, *now*."

Uubiron had a look of resignation I had rarely seen.

"There is no time, Kriss. They are going to burn the whole forest down to the ground around us!" He retrieved our contingency plan from a compartment on the wall – a bottle of pills counting enough for the camp, including the refugees.

There was a banging on the door.

"This is it for you!" a voice echoed through the barrier.

* * *

I exited the building with Uubiron and another soldier while the Kaancha and the rest of our party searched for our secondary objective – the list of operatives. We approached the bunker, Uubiron and I raising our pistols while our comrade readied the flamethrower.

A vehicle had flipped over the door from a drone blast. The three of us could remove the debris with some difficulty. As I gripped the handle, I felt a burning sensation. I reacted by pulling away and shaking my hand. I observed my palms to be undamaged. The NIB took the initiative, only to be shot through when the door was opened.

Uubiron and I leapt to the sides of the bunker as more shots exploded out of the open doorway. Logaani approached, replaced her helmet, and dropped the visor. Another soldier grabbed our fallen comrade's flamethrower, now lying next to the corpse with specks of blood shimmering like tears around its mouth.

The pair mounted their weapons on either side of the stairwell and squeezed the trigger, the immense heat evaporating the tears.

After their blaze ended, I looked down the stairwell. I saw flashes of someone tied at the wrists to the back wall. They would have surely perished in the flames, but as my sight adjusted to the darkness within, I saw no one. Logaani descended into the bunk and retrieved a miraculously undamaged portfolio from within.

The compound was consumed by the blaze around us. I looked at the documents she presented to Uubiron, attempting to read the cover, but all I could see were jumbles of symbols and ink blotches.

The world around me warped.

Uubiron and my comrades began to become twisted images of their former selves.

I found myself surrounded by looming shadows with sinister intent.

I tried to convince myself they were dark imaginations.

One form separated from the others and assured me:
We are here with you.
A chorus erupted from the others:
Remember! Remember true!

* * *

The brace holding our foes at bay was breaking. The pills had been distributed by Uubiron and all present were quiet. We had kept nothing from our followers – their enemy was an alien force dedicated to the enslavement and eventual extinction of the human race.

The children did not understand.

We had attempted to provide them with the means to defend themselves. In our exit from Instruction, we had taken a list containing the names of the future aliases our people would use to infiltrate humanity. We kept the document here in the bunker, but the ambush had left us no time to camouflage the concrete exposed above.

Two of our number had taken their pills, though the process would take roughly ten minutes. The parents cried as they shared in this suicide with their children, oblivious. I had often wondered how one could ask this of their child as the "suicide cult" narrative had become so easily believable. Some humans created cults of their own.

Their fate would be much worse as specimens of the Militia Majora.

I asked this of them now, knowing they believed me to be a god – a god that would die beside them.

Another explosion shook the door after a small charge repeatedly assaulted the barricade. The brace had been weakened, and this blast found success. Quickly, the others swallowed their pills. In a moment of hesitation, I looked to Uubiron and my comrade tackled me. The pill rolled from my hand and I saw Logaani Niih step into the bunker, crushing it under her boot.

"Sorry, Kriss." Uubiron was on top of me. "I just couldn't let you go down like this, with *them*." He looked around at the bodies littering the bunker floor, stalling as he saw the children, before shaking his head of compassionate thoughts.

"They are the enemy," he added, for his own benefit.

"You don't even know who the enemy is!" I struggled to topple him. A hand grabbed me by the hair and dragged me up the stairs.

"Well, this has been fun – but there is great work to do," came the voice of Logaani Niih. "You're lucky the Imprir has invested so much in you, or you'd end up like these animals." She held what I recognized as a conditioning device up to my eye.

"This is for your own good...," whispered Uubiron, the traitor.

"Time for a nap," Logaani added as she activated the dark cylinder. "We can blame it on the *father*, not that they know who that is...," she continued, her voice fading away.

The last thing I remember seeing before the blinding flashes of light was the fire of several flamethrowers erasing all evidence of this rebellion.

As new pathways were projected over my memories, I heard the voice again.

Who are you now?

* * *

New Mexico, present day

As the flashing lights subsided, I was overtaken with vertigo. Logaani released me and I fell to the floor. She found my condition amusing.

"Yeah, I remember my first *miid*," the Kaancha chuckled. "The reconditioning is like a bad trip, huh?" she informed me, extending a hand.

"Kaancha – Logaani – what just happened?"

"You've been brainwashed, soldier. Simple as that. Ask me, probably several times. It is a rough thing – to *remember*. Your every sense is receiving two different signals. These false memories, they can be erased by using a reversal of the pattern used –"

"No, Logaani, I don't understand why you would help me, or if you are – in fact – still my foe," I replied, taking a step back. "Why were you in both memories?"

"Not true, Kriss. I was only in *one*. Only one was a record of true events. I think you know which one."

"You didn't answer my first question."

"Let's just say your views now align with mine, as well as another old friend – Uubiron. He regrets his betrayal and convinced even *me* to feel remorse." I could hardly believe the apology.

"He even gave me the scar." She reached to grasp the right side of her neck.

"He learned the truth. He knows what we all really are – what you–,"

Logaani's speech was cut short by a loud whizzing sound. A hole had appeared in her stomach, and her suit became stained a deep red. Her expression changed to that of fear. Her eyes lost their color and her breathing stopped.

I rushed over to Logaani, ignoring the confusion overtaking me.

"Who am I?!" I shook her limp corpse in my frustration.

"*WHO AM I?!*" I screamed finally, at no one in particular. I released the Kaancha. I would have sworn that her skin shimmered, and she faded away rapidly, though I could always feel her in my hands.

I thought to observe the wound and lifted her jacket. More of the red liquid, an oily substance, had leaked out of an apparatus under Logaani's skin, unlike those used by Nemitians in their true form. I now kneeled in a pool of the substance as it dripped from my hands. It was then transferred to my face as I collapsed.

I thought the Kaancha surely dead. No form of life should be able to function after such a loss of fluids. As I succumbed to this insanity, her arm jerked, grabbing my wrist and pulling me to the floor. The second action was unintentional if I was to believe Logaani's tone.

"Damn, sorry. Can't really see all too well," she said. "Do you know who you are yet, *Mr. President*?"

I stared gawking at her before slowly shaking my head.

"Kriss, you've got to –" Her body suddenly jerked unnaturally. "Got to – cut – c– cut it out…," she stuttered, finally losing her grip on my wrist and lying motionless beside me. A ring of light flashed from within her eyes and faded until her empty iris was all that stared back at me.

As I watched the color leave her cheeks, the body before me phased in and out. I would feel its weight against my hands and then it would be gone. I thought I saw a sprite within the mess of wet tissue and oozing dark fluids, seeming to *project* the form. The fluids remained, creeping away from my fallen comrade and shimmering in the red light.

Yekk approached with her weapon drawn, lowering it only when she saw I was not a threat.

"What did you do to her!" I lifted myself up onto my shoulder. Yekk looked regrettably at the body of Logaani Niih – in a pool of the dark liquid neither of us could identify – before answering me.

"She was the enemy." Her eyes focused, and she again raised her rifle to me. "We were ordered to take you into custody once we had the weapon schematics. Where are they, Kaancha?" she said, hoping to make me more amenable to the demands of the Imprir.

This incited a rage I had never felt before.

"You don't even know who the enemy is!" I cried again as I swept one leg behind Yekk's ankle and pulling. She fell backwards, dropping her weapon. I had quickly fought off my daze, but remained in the lower position to pull off the switch. I righted myself and grabbed Yekk's rifle off the floor, knowing that if I fired, she would share the same fate as Logaani.

"Is the Militia all you know? Have you never thought to find a higher calling?"

"We are all born Militia!" she replied without a moment of hesitation. I pitied her in this moment as I readied the trigger.

I chose instead to strike her with the butt of the weapon, successfully incapacitating her.

If I was correct, Logaani's goal was to retrieve the translated weapons manuals. Somehow, Uubiron had placed her on the Sub Base-3 boarding party after they shared some moral crisis. He and the other Kaancha had known of these files, only myself left in the dark after my perceived betrayal.

My distant comrade had several changes of heart after realizing the implications of our invasion. Right now, he was the only one I could trust – the only one other than myself the Imprir could not.

This made him an ally and a target.

I had many new questions – the greatest of which was *my* purpose. I had thought I served obediently, but this black spot was usually the end of a subordinate's story.

This is for your own good, I heard repeated in Uubiron's voice.

Yekk had surely heard Logaani, yet I could not bring myself to prevent her from reporting Uubiron to the Imprir – if he had not already been captured. I worried for my comrades, surely experiencing similar trauma.

I searched the dark room for the documents we were sent to retrieve. Logaani had missed one panel behind the wall of screens used in our clandestine communication. As I removed the paneling, I saw a single flash drive. A piece of tape was placed over it and in bold lettering, all capitalized, I read in Nemitian:

KHANO INDUSTRIES

[ENGLISH TRANSLATION]

My heart beat furiously, and my anger swelled, turning my face red. With no enemy at home, the Militia Majora had become a tool for the capitalism of another world. We served a new function as a slave trader. We controlled their breeding, their opportunities, and their passions.

We controlled their very lives, many living in an altered reality, unable to compromise.

This was a venture I had decided I wanted no part in – a *choice* that was taken from me. My previous life now resumed, I retrieved the conditioning device from Logaani's hand, forced to break a few fingers. I didn't wish to leave the Kaancha, but I could hear boots scuffing the floor of the hallway adjacent.

I looked back at Yekk's unconscious form. I noted her reaction to learning of the Imprirs' plans for humanity and the disdain she wore boldly on her face at the thought of this ambush. I looked up, through the ceiling and beyond the sky above.

"I am Kriss Methius," I said to deaf ears. "Nemitian."

I retrieved a S.T.A.P. grenade from Yekk's tactical vest. I twisted the device clockwise one quarter turn – a half turn counterclockwise would activate the three second countdown.

The Sudden Target Acquisition Protocol grenade was the source of disorienting weaponry. When activated, it wails a siren likely to make human eardrums burst. A strobe of colored light induces vertigo.

Restrictions prevent human stun grenades from employing chemical warfare, but the S.T.A.P. grenade was famous for its knock out gas mixture. It was a heavily kept secret of Khano Industries.

I heard the first creak of the door and completed the activation sequence. I lobbed the grenade to angle its reflection into the hallway and heard the call to retreat.

I exited through the side door and vaulted over the balcony ledge. With the railgun under one arm, I headed across the deserted base towards the elevator.

CHAPTER XI

New Mexico, 1975

I kept to the tree line, hoping that if I ran into a soldier on perimeter watch, I would catch them unawares.

I had no trouble confirming my suspicions that this ambush was the true mission. The hangar and offices had been silent. The grass was a dark mass, flashing green as the blades were swept to face the moon by a cool breeze. I knew the elevator exit would be guarded by NIB operatives, so I headed for the only other exit – a tunnel beyond this false forest. There were not many of these bases, though they were connected underground by high-speed rail. I would exit at a terminal under the Denver airport in a little over an hour.

Plenty of time to reminisce on long forgotten memories.

It has been over half a century since my first conditioning after I realized *we* were the demons of our two realms. How far back in my lifetime would I find altered memories?

* * *

New Mexico, 1975

We had been taking Instruction on the planet and had no contact with the humanity as yet. Our leaders had recruited several disposable allies to help us establish a training facility. We passed by their graves the first time I set foot on this Earth, marked only by the freshly shoveled dirt along the small landing strip hidden in the desert of New Mexico.

While several of our number would be shipped overseas, America was chosen as our first staging ground because of its diverse inner culture and its foreign travelers and influence. Operatives had been planet side since we first discovered their realm – unbeknownst to the natives.

Perhaps we could have approached them as comrades, but the stories from the *Dark Eye* are told to all younglings on Nemi as bedtime Militia Majora propaganda.

The Old Government sent a science vessel through The Break as the first joint venture of Nemi's powers. The recent civil war had left those remaining in the Militias eager for the work they were promised when they enlisted. However, the ore processing plants on the moon had been shut down as part of the liquidation of Khano Industries.

This left a willing workforce for a new industry – the virgin asteroid mining clusters on the opposite side of The Break. The rip in space had revealed to us a realm infinite in its expanse – the newly formed Militia Majora could see opportunity among those forever stars.

Before the Old Government would allow the National CEOs to stake any territorial claims, they wished to know what kind of universe this would be and what people – if *any* life – lived there.

Communications were lost soon after the science vessel traversed a wall of asteroids that formed a barrier between the nearest star and *outer* space – a new universe. The Kaancha and crew of the *Dark Eye* were sent to recover any data that was recorded in this new realm.

The return of the science team was only the secondary objective.

They returned empty-handed, claiming *demons* had mutilated the bodies of the science crew and were holding the ship in a secure location. This played deeply on the passions of the people. Nemi's government bodies were being driven by the Temple, a religious order. The Temple Seekers came together for Communion to ask what path our god had set before us.

The Seekers lived solitary existences, separate even from one another, each meditating on their personal purpose and the purpose of *being*.

Some perished on this journey – many were the first to enter the Break, never to return.

This was an honorable sacrifice, the knowledge they gleaned from the void worth any risk.

When the Seekers exited the Communion chamber, a new enemy had been shown to them. They claimed this new realm belonged to Nemi and its resources were our right. The demons were a virus killing their hostworld and the joining of the Militias was the vaccine. The Old Government met with the Speakers – those who had journeyed long enough – to develop a plan.

The desire of the Nemitians had been the eradication of humanity. The Old Government fed this desire to recruit many younglings to fill the ranks of the newly formed Militia Majora. Seekers left their lives of solitude to lead us to war with the demonic force that threatened our existence in a holy crusade.

I now knew that this first incursion was known to humanity as the Roswell Incident. As bloodthirsty as the soldiers – alongside the religious zealots made of the Seekers – were for battle with this new foe, the Old Government and Militia Majora remained level-headed. They placed several operatives on-planet to observe the many communities.

These early records revealed a world on fire. The peoples warred amongst themselves. Paranoia was rampant between *neighbors*, fearing they were spies working for a foreign government who wished to take over their world, enforcing their way of life over all. For all the unrest this caused the world, they further provided us a canvas for our infiltration behind blacklists and the "red scare."

Our goals changed with the political environment of Nemi. Those CEOs that had been Seekers and Elders were pushed out of their positions by more entrepreneurial younglings. The Temple lost its power after the rise of the Imprir. Without the blinding light of zealotry distracting them, this new leadership found a use for these life forms as slaves. The humans – *Homo sapiens* – were livestock, animals to be bred for our eternal use.

I, along with many Nemitians, had joined the Militia Majora to hunt demons.

My position as Kaancha had allowed me the opportunity to be included in the first wave of those infiltrating this new world. I was already taking Instruction when the Old Government became the Imprir and our mission priorities had changed.

When I was recruited, I was promised what the Militias promised any youngling – adventure and excitement to fill our years working for the greater good. This meant job training, serving your time enlisted in a factory, and retiring if you survive.

Survival had become commonplace in the absence of conflict.

The Militia-Industrial complex created under the Old Government had become integral to every territory's economy. Power swelled and deflated, moving from one country to the next in an endless cycle of warfare that culled the population when jobs and resources became scarce.

When the treaty had been signed to form the State of the Three and Militia Majora, the grand scale in which wars were fought was replaced by class warfare. Soldiers no longer fought distant enemies to their homeland, instead policing the people who had become distraught and impoverished. Bringing glory to one's people was no longer the warrior's path.

Those who had served fighting each other in the Minor Militias were ready to fight again when this new enemy appeared. Those younglings who had not seen battle thought they would like an Omega pin, but victory is never the result of brute force – rather, those moves made against your enemies in the dark.

"I just landed! I was hoping to add to the graves along the strip tonight!" I groaned at my Superior.

"There will be plenty of time for that, Kriss," the Supe assured me. "They're still the enemy – the *Imprir* just seem to have found a use for them and they're not as 'gung-ho' as us here in the field."

"They've probably never seen a field. Why are we working with *any* of them?"

"To know one's enemy is to teach them defeat," said the Supe, repeating the adage now engrained into the mind of those soldiers turned spies of which I was now included. A universal *truth*, we were not surprised to find human tacticians had come to similar conclusions throughout their history.

"Sure looks like we're getting to know them," I added.

We passed by a concrete wall. Several weapons stations had been placed at a medium distance from the barrier, covered in dark red stains.

We were training ourselves to use the native weapons and until recently, acquiring targets was indiscriminate. We had pulled no subjects out of the local community in order to remain undiscovered; however, our new objective required us to infiltrate this community as a test. Some would create new identities and simply relocate to an unsuspecting property to observe those around them.

I wanted action.

I wanted to effect change if I would not take life. I wanted my being here to have *purpose*. When the call came for those of us in higher rank to serve a greater sacrifice, mine was the first name on that list.

This would be our first test and final exam. While on Earth these past few weeks, we had been preparing for battle and aggressive invasive maneuvers. None of those weeks of training mattered now as we began our Instruction in being *human*. My fellow Kaancha and a rising star recruit named Logaani Niih were moved to a closed environment hidden beneath a US Air Force base.

This was a simulated American neighborhood – a copy of one residential zone attached to a simulation of the downtown square. Some of the local communities began reporting several missing persons, but not enough to draw federal attention. As the local authorities were autonomous from the larger government – a distinction they fiercely wished to keep – they did not have the resources necessary to survey the vast forests and deserts we used as cover for our terrestrial operations.

Those we abducted were conditioned to believe we were their neighbors and treat us as such. Although we still appeared Nemitian, the process caused a subject's brain to ignore these features. Translation devices were installed behind their right ears attached to the nervous system, an addition we made to our own physiology to more quickly learn their many languages. While we too had many dialects across our world, I ached for the universal language of Nemi.

Before the cavern, accessed through the basement, was a large sterile facility where our many surgical procedures were performed to hide our true identities. The physical torture of these experiments provided the pain I would use to fuel my determination. As I shook hands, broke bread, and engaged in idle conversation with the enemy, I reminded myself this was the beginning of humanity's end.

A truer smile was never far behind.

Shortly after, however, I fell into that most common of delusions. Days – later, weeks – would pass before an Instructor would enter the camp to review our progress. If I was judged to be compromised, I would be moved to a different team whose roles were less important. This was the case with Logaani Niih, though that was to be expected from a recruit. I had served long and learned how to hide my emotions – this was especially necessary for our infiltration.

Only those whose bodies had been altered to hide our duplicity would remain on the planet, disconnected from Nemi and the Imprir. We would operate as a lone unit spread across this Earth with little direction from our leadership – this only exaggerated the importance of our obedience. Every decision would need to be prioritized to align with the Imprirs' newfound desire to rule this world rather than destroy it.

The night of our graduation, my comrades and I gathered in the simulated City Hall. In every direction, I saw only human forms and heard only human languages. When this structure was first built, we had a similar meeting establishing our goals. Standing behind the now empty podium was none other than the Named Superior of Law and the Militia Majora, daughter of Adunaan, Hi'Leng Khano.

I was uncomfortable sitting under this alien roof, the architecture requiring so much more work for the engineers than was necessary for a functional home. On Nemi, resources were scarce. Homes were small and possessions little as the necessities of life and utilities were shared burdens.

Our society was not lacking passions but a sense of *global* community restricted access to materials the Imprir deemed necessary for our infrastructure. That which was needed to replicate the angled roof could have built a two family home on Nemi.

If one needed food, there was a local pantry one could get any ingredients for any meal. There was no need for a refrigerator as attendants worked the food stalls at all hours of the day and night. If food would soon go bad, it was prepped and provided to those that required public aid.

All Nemitians know fresh food could be found after a shift change.

These were only a few of the differences between our worlds, a small sign of our people's differing priorities.

Today, amongst a sea of hairy skinned doppelgangers, the last Nemitian stepped onto the stage. Standing behind the podium, Hi'Leng Khano's red scales and golden whip glistened in the light of the many bulbs brightening the chamber – the wasted electricity was a lacking priority I find distasteful in any role.

"Today, you become human!" she began to uproarious cheers. "We have studied our enemy for thirty of their years. We have learned their weaknesses *and* their few strengths! But today, you will walk among them – you will have to call them friend! *Allow* them to exist – but know this! Their time will come!"

I had completed training in the native tongues the Imprir had surely not seen necessary for their Named Superior. This world had many languages to learn, making international business a chore only for those dedicated enough to its work. The syntax and grammar of our language was much more complex – one could say much while speaking little.

"You have each been trained to perform a specific duty. You know what your duty is, and you will perform it without supervision – of this I believe!" she chanted, saluting, placing a hand on the crest between her eyes.

"Of this I believe! We are all born Militia!" the gathered bodies replied, each saluting as trained.

"You have been molded – reshaped into the image of a demon! And I have before you now these very demons whose lives you will claim in this realm!" She stepped aside as several humans were led onto the stage with sacks over their heads.

"Logaani Niih?" the Named Commander called.

The pride of the recruits walked up and was given a portfolio containing all the identification and information she would need to move into her new life on Earth. The Named Superior looked around and, dissatisfied in her search, pulled her personal rad pistol out of its holster. She pointed at the first human and the soldier who led them onto the stage removed their hood. It was a woman with short, dark hair.

"Your last task is to get rid of this evidence." She handed the pistol to Logaani. The recruit looked at her superior, then looked out to the crowd, meeting my gaze. She turned to the woman, kneeling with her mouth bound, unable to cry out.

She fired the weapon and nothing was left but a piling ash and bits of bone falling through the haze of burgundy vapors.

There was a moment of silence, perhaps because this was the first casualty of this war many of us had seen, though it was not long before many cheers filled the hall.

Another name was called, and another soldier approached the podium. There was not as much of a delay before they fired the weapon and read the name on the portfolio aloud.

"Bennewitz – yes, I've done my research."

"Kriss Methius?" the Superior called. I stood, and many eyes turned to follow me as I approached the podium. I read the name on my portfolio for all to hear.

"Oscar Orwell," I said plainly.

I was a Kaancha in the Militia Majora – a high station to hold in our society. My responsibility for this mission was greater than most. I was trained for every situation and in the political playmaking of the world's major players. My position would be highly visible to the population of this country and, if all went according to plan, I would soon have the attention of the world.

I turned to the life before me, fear widening their eyes and tears soaking the gag in their mouth.

I squeezed the trigger.

I shook the hand of the Named Superior and a conditioning device was held level with my eyes.

We were each to be conditioned to believe we had lived the life of the one we were replacing. It would be easier for us to fill our roles –

* * *

This was the story I had been forced to believe, the first sign a memory of *willing* conditioning. What truly happened was that I had refused to take the life of Oscar Orwell. Hi'Leng Khano was not present – the ceremony was presided over by High Inquisitor Hi'Ran Ahket. They had been forced to alter several memories after I defected to Vietnam to keep me in line of which this was the first.

Remembering who I was would be a long journey.

Those of us destined for leadership positions had to begin our infiltration among the youth of humanity such that here was a documented identity that could not be refuted. I was educated and passed my exams at university. I fought and bled for American values in wars passed. For this to be possible, the identities we Kaancha assumed were that of seventeen- and eighteen-year-old students.

Their families would disappear in tragic accidents we would heroically overcome, immediately placing us in the spotlight. At graduation, I refused to take this youngling's life. While I argued with the Inquisitor, Logaani Niih took the shot and claimed she simply wanted to move things along. Much was left unsaid, but I was allowed to continue in my position undergoing my second conditioning.

* * *

I had been riding in the rail car for an hour.

I felt the car slowing and prepared to finish my escape.

CHAPTER XII

HOPE Campus

Below the Denver airport was a checkpoint between underground bases in Dulce, New Mexico, and Cheyenne Mountain in Colorado. The station was not as expansive as these natural caverns, hidden behind another false barrier in some back hallway. I assured myself there were no witnesses as I walked through the wall and found my way to the boarding gates.

I was unbothered by passersby as I exited.

The sun was beginning its ascent. I quickly stowed away in a dusty truck bed with a tarp to cover my form. My chauffeur unknown and myself exhausted, I allowed rest to overcome me. The sound of a violent horn section woke me to the tune of downtown traffic.

I attempted to remain hidden as I peaked out from under the tarp, immediately lifted on the wind.

"Hey, what do you think you're doing?" a voice called out from a bearded face, now eyeing me out of the driver side window.

I frantically searched for a lane to make my exit but found we were in gridlock.

"I didn't mean to spook ya, you just gave *me* a scare. Look, I don't know where you're coming from, but I can take you to a place that'll get you straightened out."

Not in any position to question this charity, I quietly joined him in the passenger seat.

"Might as well tell me your name. We'll be here awhile." He gave his own horn a couple honks.

"It's...Kriss," I told him truthfully.

Traffic moved again and as we navigated the streets of the metropolis, I noted the many abandoned homes now designated construction zones, some already torn down for parking garages for new skyscrapers expanding from the inner city. Since the early end of prohibition on marijuana, Denver was one of the first areas to see tax revenues from this new economy. However, instead of investing in improving what communities needed help, they invested in aesthetics – more important to look progressive as opposed to living so.

These expansions were driving out communities of low-income housing. This only worsened during the pandemic, with the closing of many local businesses, increasing the already inflated population of homeless citizens.

I turned to my companion, whose incessant rambling I appreciated. Andrew asked little of me. He filled our time together with stories of his family. He was most proud of his eldest son, a recent graduate.

"He's a good kid," he added as we approached a large brick building. "I only hope the jobs are there for him...well, anyway, here we are! If you need anything, just ask for Mrs. Cleo."

I thanked Phil for his kindness, waving back at me as he drove away from the homeless shelter.

One would believe they were invisible. The lack of attention I was given – several pedestrians nearly bumped into me. As I looked down, I saw my clothes were covered in patches of dirt. Twigs had become entwined in some threads.

No one wishes to be reminded of their own mistakes or those of society as a whole. Many with psychological disorders are abandoned, incapable of affording the care that should be provided to them. Absent parents never reach out to those they have left for fear of renewing the pain they inflicted and being forced to admit their failings.

If they only knew that many of those forgotten by society or aging out of a broken foster system of public care would end up homeless.

More privileged eyes would avoid dark alleys or roll up their window for a sign-holding citizen at an intersection. By ignoring the problem, they convince themselves that there is nothing more they can do. In a city priding itself on progressive ideals, it was no surprise that the lack of public aid saw many others living on the streets of Denver, and I may soon join them.

New predators have evolved alongside society. As they hoard wealth and abuse the power that follows to restrict others from achieving it, they force many into a more primitive existence of survival over choice. They offered meager supplement in the form of understaffed shelters and public kitchens.

As I exited the kind man's vehicle, I saw a line forming of my now-fellows. I quickly retrieved the railgun from under the tarp and enclosed it in a loose trash bag, and slung it over my shoulder.

A shelter was about to open their kitchen, feeding these needing souls throughout the day. The food was prepared and served by volunteers, many who had lived on the other side of the line as well. Some served from the back of their van or truck, providing for those less fortunate who could not travel.

I approached the shelter, leaving the rail gun behind a dumpster in the adjacent alley and stashing the flash drive in my jumpsuit. I was glad to find the line short – I expected the longer I remained at *any* location, the more imminent my discovery would become. As I took my place, I noticed a woman and a child appear on the sidewalk across the street. They were followed by a couple – this particular alley was home to many.

My hunger was audible.

The woman and her child wore clothes that were of recent fashion, perhaps of the last season. Regardless of their prior circumstance, they seemed to be new additions to this neighborhood. Many might judge a parent living on the street with their child, but those are usually people privileged enough to know choice.

Some were *already* homeless or addicts, many attempting to recover their lives. While there are options when one is not prepared to be a parent – none of which is any greater or lesser than the other – the state cannot argue whether a parent loves their child based on their inability to support themselves.

"It's the bureaucracy of it all," I heard a man wearing a large jacket at the front of the line say to a volunteer. I assumed he was a veteran, as indicated by the easily recognizable fatigue green and Air Force Staff Sergeant patch on the shoulder.

"What do you mean?" the woman asked as she passed the man a plate. He stepped aside to allow others to move down the line, expanding his audience and sharing his insight.

"I mean the HUD office – the government – they make so many hoops for you to jump through. It's bad enough that we lost our jobs because no one could keep open! Now all you can do is work distributing for some big company that could survive the pandemic! Or in a factory, and those jobs are going to the *robots*!"

"He's right, you know," I heard a voice add from behind me. I turned to see the mother and child were next in line.

"We haven't been here long, but everyone seems to know each other." She adjusted her scarf. I saw the couple that had followed them was further down the line, happily talking with friends rather than take the next spot.

"But...I haven't seen *you* before," she pressed. "Do you...need any help?" I fought back a tear at the thought of someone *here* offering me assistance, a testament to the capacities of human compassion.

"No...no, I'm OK," was all I could manage. She offered I may sit with them, introducing herself, Renee, and her daughter, Felicia. I accepted and as we approached the prep table, the man was still at the volunteer's side, not a single bite taken.

"You know they have different classifications for 'homeless?'"

"In fact, I do, Charlie," she replied. "The HOPE Campus is tertiary, longer term free housing, and you would be in too if there were more rooms."

"Yeah, well, the only charity I want is the government's! What they owe!" None disagreed with his agitation. We received our plates and as I attempted to pass by unnoticed, he grabbed me by the shoulder. I dropped my plate as he turned me about facing him.

"I know you. How do I know you?" he asked with a wild look in his eyes. I feared I had been recognized, although to no surprise.

"Charlie, that's enough!" Another volunteer had heard the commotion and appeared to calm the veteran.

"Lucy, I know this man! I know him!" He would not let go.

"Sir, please." I grabbed the man's wrist. I realized now that he did not know my identity. He was most likely suffering a manic episode from some form of post-traumatic stress disorder.

Lucy approached him calmly while everyone else put space between themselves and the two of us. Charlie turned to her, letting me go, and walked away with Lucy to sit at a table just the two of them.

"He doesn't mean it," the volunteer insisted.

"You don't have to apologize for him, I agree with everything he said." She stared at me for a moment, then passed me a fresh plate. I sat with Renee and her daughter, who had already begun eating.

I was consumed with my own suppressed thoughts, hidden from me for thirty years. I did not know the repercussions of reconditioning. It was a rare procedure, only performed if there were dire side effects to the initial conditioning process – memories that could not be hidden, causing a psychotic break.

"Do you have anywhere to stay?" Renee asked, reading the concern on my face.

"I...do not." I was reluctant to make any complaints to the mother.

"You should talk to Lucy and Mrs. Cleo – she served us the food, that's Lucy's wife. We stayed with them when we first arrived," she informed me, scratching her daughter's hair. They both smiled at each other.

"But I saw you in the alley." I inquired.

"We stay on campus now. I didn't know then, but we were classified as secondary. By staying with family and then with Mrs. Cleo, I couldn't apply for rehousing," she began. "I worked my whole life to run my own restaurant. My husband and I had Felicia just before the pandemic – and the business was new. We had to close before we could even open and I lost all of my investments. My husband had to support us alone, but he wasn't getting the hours he needed either," she said, pulling her daughter's head to her chest and holding back tears.

"He was angry I couldn't help. I couldn't go back to serving – what jobs were left were taken." She ran her fingers through the child's hair several times. "And now we're here, and we're relatively safe her at the shelter." She looked back up at me. I noticed bruises partially hidden by her scarf to match my own.

"Mrs. Cleo, she said she would lie and say we weren't staying with them, but they do so much to help everyone around here already I couldn't put them at risk. HUD is still a government agency, and it would be a federal crime, you know, just to lie on the document," she informed me. I played dumb to any knowledge of law, though I could see no judge prosecuting such a charge.

I simply nodded my head.

"It takes time – three to five days to receive our application, eight to ten days to process, and who knows how long before we are assigned an address. I didn't know a lot of things. Did you know we can still vote? The shelter can be an address, or even a park or street corner. Mrs. Cleopatra told me that too," she said, again praising the volunteer. "I didn't vote for Orwell, but it still counted for something."

"The most basic right, choice. Even if it is limited," I said. Renee found this statement humorous and chuckled. When she opened her eyes, she waved at Lucy, who was now approaching the table.

"Welcome to HOPE Campus, I'm Lucy – Lucy Peterson-Chambers. Don't mind Charlie, it's not his fault. Same goes for many around here." She extended a hand I quickly shook. I continued to keep my face down.

"Do you have a story to tell?" she asked, adding, "Say, that's a strange outfit."

"He's looking for a place to stay tonight, isn't that right?" Renee inquired of Lucy on my behalf.

"I suppose if it's OK with *bae*. You can wash up here. Are you hurt? You've got a little blood there," she observed. I knew it was the strange liquid that had bled out of Logaani Niih and ached to be clean – the jumpsuit no longer served its purpose as a stealth device.

"That would be great. Can I possibly get a change of clothes as well?" I remembered the many company and Militia patches covering my jumpsuit.

"We have laundry in the back. You could find something there," Lucy replied. "Just be sure to rewash it. My wife and I will be here a few more hours. You find an outfit and wait here with Renee and Felicia. We'll come find you," she said, leaving the table to assist in the food line. After a few steps, she turned and asked, "What did you say your name was?"

"Kriss," I stated without hesitation.

"Well, welcome to town, *Chris.*"

I sat with Renee and Felicia for another hour and wished them well in their application. I entered the bathroom and looked in a mirror for the first time. I had a couple of bruises and dirt on my face. A few flakes of dried blood could be seen if one ignored my swollen nostrils. A painful disguise, but it would serve me for now.

I hoped to accept a peaceful night's rest from the Peterson-Chambers. I would leave in the night to find someone I could deliver the weapon schematics to. If humanity was aware we are among them, they would surely have formed a resistance and I believed they would be looking for me.

I knew that the Imprir would learn of my newest betrayal and the true identity of President Oscar Orwell would be made public, leaked to news outlets from "top White House sources."

I washed my hands and exited the bathroom. I found a room full of clothing donations and suitable dress: a pair of blue jeans, work boots, a white t-shirt, and a hooded pullover.

Secure enough, I pocketed the flash drive containing the schematics.

I did this small load of laundry and changed. I disassembled the rail-gun and wrapped the parts in the jumpsuit. I tossed it all in a large black trash bag and returned to the dumpster behind the building, disposing of any evidence of my identity. I heard a scuffle behind me, but it was only a dog digging through the trash.

I was paranoid because of my recent reconditioning and rogue status. I found the table I had shared with Renee and her daughter now occupied as many more hungry souls had gathered while I changed.

There are more food pantries and shelters like this one every day. It was my belief that the government could do more to provide its citizens the means to become productive members of society – if for a second time. It was a position easily championed by leaders and those who would in turn champion the cause to friends – all of whom would do nothing beyond talk.

The homeless living on the streets of the nation's capitol numbered twice that of any other city.

The disparagement between the classes would always be beyond the capability of these continually defunded systems. The resources were available, but the priorities of public service are lost under business interests and opportunities of advancement. These voices remained un-heard – these members of American society truly the ones without representation.

Had I only been truly awake.

CHAPTER XIII

Reboot//NIGHTMARE

Lucy and her wife – Mrs. Cleo to those at the shelter but she preferred Cleopatra – served food throughout the day. When they were finished volunteering, Lucy drove us to their apartment and Cleopatra left with their shared van for work – she managed the third-shift at a factory while Lucy was a nurse. We used the elevator, surely a force of habit for Lucy, although the opposite of my own.

I preferred the stairs.

When we entered the apartment, Lucy explained that while she worked longer hours than her partner, she had requested to have a day off and volunteer at the shelter with Cleopatra.

"You can stay here tonight – on the couch. Don't mind the cats," she stated, stroking a black cat sitting on her lap. "His name is Jar-Jar – because of the sound he makes! Or is this Beanx? *He* likes to be left alone, but they're so hard to tell apart!"

I did not see another, though they may be hiding. The cat I *could* see hissed as I reached out to pet it and hopped away.

It had been found that animals could sense our false identity.

I thanked Lucy and prepared to take another shower. I kept my new clothes with me and hid the flash drive under the jacket. I washed off the fluids of Logaani Niih, resembling my own blood after treating it to appear human. The effects had expired after my night on the Moon

Base, but the translucent patches had gone unnoticed, their shimmer lost in smudges of dirt.

I redressed and pocketed the schematics. I recognized the smell of a burning marijuana cigarette through the bathroom door and found Lucy sitting on the couch watching television.

"Hope you don't mind." She dragged on the smoking cone. "If it was just a fixation, we'd vape – but we share a prescription," she added with a wink.

"*More of these attacks are expected as the government remains silent. No word from the White House – kzzt,*" came a report before she changed stations.

"*Coming up in your ten o'clock power block – is your neighbor an alien? – kzzt,*"

"*Tonight on Kimmel! Dennis Rodman disputes claims that he – kzzt,*"

"I can't thank you enough," I said, taking a seat on the couch next to my benefactor.

"We're happy to do it. We consider ourselves privileged even if some wouldn't call it much."

I remained silent, planning my escape and having no reason to familiarize myself with the couple – regardless of their hospitality.

"You know, Charlie knows Cleopatra from overseas. Personally, I think that's why he refuses to stay here, but he doesn't have a plan like Renee. Do you have a plan?" she asked, turning off the television.

I did not know how to answer the person before me. I was being assaulted with the thoughts and opinions of a buried identity – many of which opposed the actions I had taken. My insubordination began with a desire to save one life, but I had been forced to make decisions that cost many.

I had a plan, not that I could share its details with my companion.

"I'm traveling. I can leave in the morning."

"You *are* welcome here, Chris. Now, how long? That's not up to me – she's the boss."

It was a small apartment, the kitchen attached to the living room divided by a barrier between the carpet and linoleum. There were two chairs around a single table making a dining area, all sitting on a large rug. In the corner was a desk with a laptop. The living room had the couch and one recliner facing a rickety entertainment center topped with small fantasy figurines.

"Do you play?" Lucy asked as I stared at the characters. To my confused look, she replied, "Guess not. Me either – that's her nerd shit."

The bathroom connected to this room and a bedroom through a door I had seen when I showered. I heard a cat *MEOW* and the second bedroom door creaked open. Through the crack, I saw what looked like a nursery. A crib was resting against wall paper of a pastel blue.

"Do you have children?" I asked. Lucy glanced in the room's direction, promptly got up, and closed the door. She came back to the couch and was quiet for a moment.

"No, we do not," she said with finality.

"I'm sorry, I didn't mean to –"

"No, no, it's fine. We should probably get rid of all that stuff anyway. We got excited, is all. We realized how much we wanted it. Then...."

I wished I could take back my question. I could see that I was causing the woman to relive what was the loss of their child. Her kindness for me – a complete and duplicitous stranger – amazed me. That she would confide in me this tragedy racked me with guilt. I hoped that this was her catharsis and quietly listened.

"We were on a trip, visiting friends. It was a long drive, and they say you shouldn't do that when you're pregnant, you know?" she told me as she stood up and walked over to the sink. There were two cups, two forks, and one bowl, but she washed them three times while she continued talking. "They were *my* friends, so we stayed the night, took the long drive home, and went to the ER. But...we should have left sooner. I guess we were looking for reasons to blame ourselves – blame something. But now we know there were complications with her vaccination – and she wasn't the only one. I'm a nurse, so we got the first wave. But we couldn't blame that either – if it hadn't been rushed, if they –," she stammered.

"I'm sure she doesn't blame you," I attempted to console her. She grabbed the edges of the sink and the dishes clattered on the metal.

"And now they're telling us the government is full of impostors or something," she said, motioning to the silent television. "Like they put chips in us – that they've *been* messing with us. I'd love to get my hands on them, *love* to blame them – but that wouldn't do anyone any good."

"I'm...sorry," I repeated.

"It's no problem, really. There's no way you could have known," she assured me before turning away and looking through the wall in front of her. "But, since you do, let's avoid the subject when my girl gets home. She already fights so much – she's the strongest person I know. She had lots of people to support her – she carried our baby. I would never argue that, but...I hurt too, you know?"

She walked over to the bedroom, stepping in and slamming the door. After a few moments, the door creaked open and her head appeared.

"Feel free to stream something, between our friends we got all the things. I think I'll just find a book, let you have the TV." She paused for a moment, perhaps regretting offering me a place to stay.

She closed the door, and I was alone.

I thought about my host's loss. The vaccinations Lucy spoke of were in response to the same natural pandemic that closed Renee's restaurant. We were not in power yet, though we would not cause such an event – a shepherd does not wish harm on their flock. Many countries could handle the spread of the virus through quarantine and public awareness, which would have given pharmaceutical companies a more appropriate time-scale to build a vaccine.

In the United States, the expression of one's freedom was thought to be so important that one would risk the lives of other American citizens for no reason other than to prove a point. The religion of conspiracy we had created caused a distrust in both political parties. Our original aim was extinction, and we convinced them government programs and public aid were a tool of the enemy of the day.

When a new viral strain appears, a new vaccine is developed. After careful testing, inoculations can be distributed to prevent the spread and eventual evolution of an outbreak. This has been a standard function of the Moloke Institute on Nemi for generations. Most vaccines are safe to distribute even in our birthing pools.

We spread lies that these vaccines had sinister intent. The connection to developmental disorders was real but highly exaggerated – there was a risk in any medical treatment. Following technological progress, the narrative changed. Today the government wished to track its citizens every move by injecting nanotechnology into their bloodstream, another bit of Nemitian truth we shared to incite the peoples fear.

Diseases thought extinct returned and new viruses rampaged through communities.

We birthed a distrust in science itself. Ideas disproven long before our arrival resurfaced: the flat earth view of reality and race supremacy ideologies. These movements proved difficult to manage when education is made the enemy, but helped to discredit the established leadership.

We had convinced the American zeitgeist of the need for change. After Vietnam, the role of Ambassador was replaced by elevated military officers we replaced in the field. Military thinking led the world into many conflicts as we established control.

Unexpected was the rise of personalities to the status of world leaders – that their strongly held opinions and lack of professionalism would *attract* voters.

These personalities were quickly shown to be inadequate. The people wished for proper leadership. I stepped in to fill that role while continuing the tradition of misinformation.

The memories I had from my years serving as President of the United States were conflicted. To remain benevolent publicly, I pushed for change that benefitted the American people – but was that truly me? These memories were tainted by the knowledge that I was not in control. Every positive was made to obscure some negative.

I was glad to be free to choose what I thought was best for them by providing the means to defend themselves.

I stretched my body the length of the couch. I removed the hoodie and wrapped the conditioning device within, placing it between the couch frame and the cushion. I finally fell asleep, feeling comfortable and safe. I knew this was exactly where I was supposed to be.

* * *

I found myself in a medical office on Nemi. As I looked down, I saw my true form.

Blue scales swirled with indigo, my clawed hands speckled with white. A common pattern, but I was ecstatic to be back in my body. I looked to my side and saw a familiar face – that of my wife, Ngema.

She was nothing short of extraordinary, her tall stature included. The women of Nemi were taller than the men and she boasted three units on me still. She wore a canvas of solid black – rarely was there no pattern, especially after the shedding. Like the High Inquisitor, she suffered from Tympania Dysmorphia, her skin scale-less.

I thought it made her more beautiful.

I realized I was lucid in a memory and not a good one.

"[No – we have no reason to believe the *Tympanodrome*,]" said the doctor – in Nemitian – on the other side of the desk, "[had any effect on the egg. When the egg hatched, the youngling's lungs didn't activate.]"

"[What are you saying?]" Ngema shouted, showing her face for a moment from behind her clasped claws. As I desired to look away, I noticed the state of the furniture and photos on the walls was shifting.

"[Your youngling –]" the doctor started.

"[K'Haan,]" I said quietly. "[Their name – was K'Haan.]" I spoke uncontrollably, following the script of this memory.

I tried to look at the documents on the desk. Though they were upside down, I could tell it was gibberish. The last memory I had was lying on the couch in the apartment. I realized I was reliving the moment in a dream.

"[...the staff was too late, the child drowned in the amniotic,]" the doctor finished. As I turned to Ngema, I saw her now on the bed of our home, holding a pillow tightly while she cried terribly.

"[Why...why...,]" she whimpered, squeezing the pillow repeatedly and soaking it with her tears.

I continued attempting to break out of the dream, observing the vague resemblance to our furniture at home and our clothes. I looked towards the mirror on our dresser, but within the frame I saw a complete emptiness in place of my reflection. As I gazed into this void, I felt as though I was falling. The feeling passed in a moment, but as I turned back to my tortured partner, she was no longer crying.

She was sitting on her knees with her head cocked to one side playfully. She turned her head to the door before speaking.

"You're awake, but you're not," she said in English and a voice that was not her own. It sounded distorted, as if run through a filter.

"Is this not a dream?" I asked. I had lucid dreams before, with intent, but I had never been prompted by any figment of my consciousness.

"Dream, memory – memory, dream. Sure, but dreams can be real, too. Here, there, somewhere else," came the voice from Ngema's lips. The form pointed behind me at the door to our bedroom. It flung itself open, slamming loudly against the dresser and breaking the mirror.

I was blinded by an assault of wind and heat. My eyes burned from the brightness of the surface of the sun bounded by the doorway. I was pulled into the light and landed hard on an invisible surface. My eyes no longer burning, I looked around and saw nothing.

Nothing at all.

I stood and held out my arms, believing my balance to be hindered. I appeared human again and began wandering into the light.

"Hello?"

"Ngema!"

There was no echo.

I wandered for what felt like hours. I doubted I was in a dream. I imagined I died in my sleep – possibly discovered by my host. I walked in many directions, finding the same lack of result. Eventually, I sat, crossed my legs, and closed my eyes.

I believed I had found the What-Comes-After.

"Didn't mean to keep you waiting," came the voice immediately upon this acceptance. I opened my eyes and saw no form.

"I'm dreaming, I'm dreaming," I chanted.

"You're dreaming, you're dreaming," the voice parodied. "You know, you could be wasting time." I laughed out loud.

"I've been searching here for...*some*thing for hours!"

"Hours – years – I forget what it feels like," the voice replied.

"Where are you?!" I slammed my fist on the invisible floor.

"Well, where do you *think* I am?"

I was exasperated. I threw my head to the left, my eyes searching wildly.

I looked to the right and saw a blur of motion. I thought someone had been there and dashed away at the last moment. I stared in the direction I deduced they would run, and the blur became a dark whirlwind.

"What do you see?" came the voice over the winds. As it spoke, it took on a form.

It was human.

"What do you *hear*?" The voice was becoming more refined. The swirling winds slowed and dissipated, revealing a slim form with short, dark hair. They were wearing a bomber jacket.

It was the form of Logaani Niih. The entity looked down and chuckled.

"I see, a more recent memory – more *nightmares*," it spoke, with Logaani's voice.

"Why have you brought me *no*where?"

The entity looked puzzled.

"Not nowhere – *every*where," it said, waving a hand. At the edge of this motion, as if a slice cut from an unseen cable, an image appeared. A swirling light formed a border to our view.

I saw myself asleep on the couch.

The hand waved, and this image was replaced by that of me sleeping on the recliner. The hand waved – again I saw myself on the couch. Two white cats slept on the recliner, the foot rest stretched out. Faster and faster, the hand motioned with a buzz. The hand stopped, revealing a group of humans gathering in the woods. They were armed and readying their weapons.

"*Almost* there!" One more motion of the hand, one more change of scene.

"This is *every*where," the voice said, though it was no longer a mysterious and incomprehensible statement. "You can go *any*where. Nod if you understand."

I nodded my head. All Nemitians are educated in several theories of reality through courses in mathematics, the sciences, and philosophy. The many-worlds theorem was one human analog. Some believed our consciousness was simply a vehicle for Salvation to experience its creation.

None negated another, but many argued which was more true. As I pondered these thoughts, the form before me nodded its head.

"You're getting there, but for now, you have to *wake up.*"

"You brought me here!"

"Just go back through the door," came many voices, booming throughout the space. I covered my ears.

I realized the voices inside my mind could not be driven away. Their volume challenged my equilibrium, and I struggled to raise myself from the ground.

A few steps before me, a door raised out of the nothingness. As it rose, it shook with a loud knocking cutting through the voices. It opened, and I saw the dark emptiness I had seen in the mirror, a sharp contrast to the infinite light around me.

I dove forward.

CHAPTER XIV

PlanetWatchers

I awoke after Lucy had answered a knocking at the door. I could not hear them and as I looked over the back of the couch and I could not identify the shadowy figure she spoke with. I did, however, recognize the barrel of the weapon I had disposed of behind the shelter as they stepped inside. I twisted away and pretended to be asleep.

I had no time to prepare any defense.

"Wake up!" I heard as I was shaken. Lucy threw me back onto the couch and as my eyes adjusted, I recognized Charlie at the door holding the railgun. It was pointed in my direction.

"This is a gun, right?" Charlie asked, already knowing the answer. "Well, I know one when I see one!"

"I don't know what that is," I lied, poorly.

"I followed you, I knew I knew you, I knew! It's the President, Lucy, look at 'im! He's one of them!"

"I guess I couldn't see it, but now that you're all cleaned up...."

She shot a glance at the television, and I turned to see my face. I had waited too long, and the Imprir had exposed my identity while keeping themselves well-hidden. I also saw the face of Adebowale Ikande – my comrade, Uubiron, had been discovered.

Without we two, The Imprir did not have control over the nuclear arsenals of the planet. We were both targets, but I was glad to see he remained at large, somehow alerted to Logaani Niih's admission.

"What is this?" Lucy held up the conditioning device. "What does it do?"

I was afraid. I wondered if this was how the crew of the *Dark Eye* felt when they saw humans for the first time. In the bedtime story, many details were omitted, sure to scare any youngling.

The official Kaancha Report described the brutalization of the corpses of the science team.

"Please – I do not wish to harm you!" I pleaded, holding out my hands.

"Let me guess – you come in *peace*," said Charlie. "I've played that role – I know that lie!"

"I'll prove my sincerity! I want to help you *all*!" I saw a bowl left out on the table with one fork.

I grabbed the fork and as I held it before them, Charlie shouted down the barrel of Yekk's weapon. I held out my open hand and cut the skin deeply.

A shimmering patch appeared, quickly dripping onto the floor.

"I am not one of you," I stated, dropping the fork.

"Well, we already *knew* that!" Charlie laughed.

"I've been here a long time – I've done a lot of things I'm not proud of! But I was not in control! I tried to stop them, but what *that* is," I informed them, pointing at the conditioning device, "is evil."

"Why? Why are you here?"

I told them of The Break and the crew of the *Dark Eye*. I told them of my recruitment into the Militia Majora and our initial infiltration. I told them of the Imprir – their plan to usurp the resources of this world and enslave humanity.

"Are there more of you?" asked Lucy. "Or – I mean – are there any other species?"

"No, you are the only other life we have found." My host's eyes suddenly grew wide and fierce as she leapt towards me.

"Did you make the vaccine?!" Lucy grabbed me again.

"No! No, I swear!" I pleaded. "We did nothing to the vaccine – we didn't want a cure! Believe me, I know this pain!"

Lucy stared deeply into my eyes. Her pupils were shaking as they search vehemently for some falsehood in my confessions. She released me and sat at the table in the dining area.

"You're saying this device, that it, what, *brainwashed*, you?" Lucy asked, holding the conditioning device loosely. "What all have they made you do?"

"It alters memories. It can change what you believed your own choices to be, changing who you are and how you will act in the future. It conditions you to assume a new identity."

"And whose identity is this?" Charlie asked.

I told them of Oscar Orwell.

"You're monsters, all of you! And Travis Walton? I've seen him speak! They made a movie about his abduction!"

"I think...this *is* they," Lucy added.

"We controlled much of the media, yes, to instill a natural fear of anything alien. We thought it would make you more easily subdued when the Imprir revealed themselves to the world as its new supreme leadership."

"What about all the good stuff? Mr. Spock and all that?"

"That was before us – and we *still* don't control everything. I took the name Oscar Orwell in 1975 – as an eighteen-year-old boy – and my reconditioning failed. It was another decade before I was active and another 20 years after that before I could really affect any change."

"You should have done more research," said Charlie. He lowered the weapon. "This was my dad's jacket. He served in Vietnam and he came back from the jungle with crazy ideas. And the government threw him aside – *you* threw him aside to cover your tracks!" he yelled, wetting my face in his anger.

"Charlie, maybe sit down...," Lucy suggested over her friend's hyperventilating breaths. Charlie held out a hand to silence our host. He caught his breath after a few more moments.

"But he said – my dad said he heard there were good ones too. My mom believed him, 'til the day she died. She supported him – us – and we still ended up here, trying to get help!" He propped the railgun by the door.

"He taught me a lot – he knew this day was coming," he said, sitting next to me on the couch. "And that day is soon, yeah?"

I simply nodded.

"And you wanna help us?"

I nodded. The two looked at each other.

"Our control over the nations' nuclear codes *was* the power we held over you, but now...," I said, motioning towards the continuing news coverage. "I have allies – *you* have allies! Some have already given their lives this day!"

"Then I hope you have something to show for it."

I retrieved the flash drive from my pocket.

"This. We need to get this to someone who knows what to do with it. This could be your *salvation*."

"Show us what you're worth," Lucy ordered, turning the laptop towards me.

I plugged the drive into the proper port and attempted to access it. There were several levels of encryption that this personal laptop was incapable of bypassing.

We could not view the names of any files or folders contained within.

"What is supposed to be here, *Mr. President*?" Charlie chided.

"Please, it's...it's just Kriss."

"What's on the drive?" Lucy parroted the question.

"They are schematics. How to use the resources available here, on Earth, to power our beam technology and –," I paused as I realized this was not the resistance I had been hoping to find.

"And *what*?"

"How to build the weapons yourselves." I stepped away from the laptop and table. "But I don't know who to bring this to – who might have the capability to open it."

"I might," Charlie chimed in. Lucy and I both looked at him in true disbelief.

"Charlie, what are you talking about?" Lucy tapped her foot, now the impatient one. Charlie sat down at the table and pulled the laptop in front of him.

"This consumed my father's life – *and* mine." He began typing quickly. "There really is a large community of believers – true believers!" He turned the screen to face Lucy and I. "I found this group a few years ago, when all of those metal monoliths were popping up? They were all hoaxes, but on *this* page..." he added.

It was a social media page for a group called "PLANETWATCH-ERS," a private forum for the sharing of close encounters and other extraterrestrial experiences.

"These people are *hardcore*." Lucy scrolled through the first few posts. Many read like a bad tabloid article – stories of abductions and reptilian visitors.

"Yeah, yeah, I know, but – but –" Charlie turned the laptop back to navigate through the many posts.

"Here!" he cheered, showing us a photo of one of these monoliths. "This is right outside of town – they showed up everywhere, but this page only shared a few. I went out there and they were having some sort of meeting. They let me sit in and *they* said you weren't real! You, the President of China, Russia – all of you, *aliens*!"

"They were right," I admitted.

"And they've been *preparing*. Lucy, you could contact them, see if they would meet us!"

Lucy was looking out the window as the sky was changing color.

"I'll need to grab CC soon. She's the boss." Lucy and Charlie agreed they would give me the benefit of the doubt, but the veteran quietly eyed me after Lucy left to pick up her wife.

We waited as more reports came in – tales of the American response. They did not pull more of their leaders into the streets, nor did the military assume control. For what could be seen in the streets, they had been consumed by complete anarchy. Grudges were closed as mobs purged their neighborhoods of undesirables, falsely claiming they were alien.

In the end, they would all bleed red.

The news had not claimed this neighborhood, but Lucy and Cleopatra arrived, having escaped an active intersection. Lucy had received many calls from the hospital but made no stops. Cleopatra believed Charlie as well, and we attempted to contact the PlanetWatchers.

While the world had gone mad, an admin posted a single word: RECRUITING

"I wouldn't guess this is *not* a real person," Lucy said as she messaged the social media profile hosting the PlanetWatchers page. "I mean, look at the name."

"Isaac Asimov...," I read aloud.

"Like, the writer?" Cleopatra inquired, turning to me. "Science fiction – was he one of *yours*?" She took a drag from her cigarette.

"You see how easily prone to conspiracy you are," I joked back. These would be my comrades in the coming fight, and I hoped they would consider me the same.

I was relieved to hear a chuckle.

"They responded."

They had sent a link that loaded several web pages before opening a single chat box:

q: lucy. do you believe

"I don't know how to respond."

"I can handle this, at least," Charlie said, sitting at the laptop and typing:

guest1: we believe

q: have you always believed

Charlie paused and scratched his head. He repeated:

guest1: we believe

After several minutes, there was no answer.

"Tell them we have sensitive information we can't access ourselves," I suggested. Charlie began typing again:

guest1: we need help. we cant open source file
guest1: khano-xx49729g

This garnered the desired response. The third and final line from our benefactor read:

q: 12062020

The page closed itself.

"What do you think, Charlie?"

"I don't know. It looks like a date," he responded.

"Check the PLANETWATCHERS page," Cleopatra suggested.

A quick search found an article posted about a metal monolith appearing in a resident's front yard. Several comments followed dated the same day – random discussion relevant to the time, I recalled.

The last comment was dated two years after the fact. The post had been inactive just as long. It linked to a new article concerning the disappearance of the monolith. The man who owned the property admitted to making it himself to join in on the fun. One comment was made by an anonymous user. It contained a new set of numbers:

39.76/104.41

"Coordinates," Charlie correctly identified.

"Just outside of town, it looks like," I added.

The group assembled around the table turned to me.

"How do you know that?"

"Well, I know the coordinates for the old Watkins Space Port," I informed them. "It would have been our landing strip," I added too late. They all exchanged suspicious glances before returning to the task of locating the monolith.

It was apparent my humor was missed.

"Maybe...forty-five minutes to an hour. There's the Space Port...And right here, Bennett...outside of Watkins! The coordinates will take us to the chapel cemetery!" sleuthed Cleopatra.

"What is the plan here?" I inquired. "Believe me when I tell you we investigated every cattle mutilation or abduction claim – that wasn't us. Most of these groups are full of nuts."

"All we nuts need is proof to be taken seriously – all *they* need is you."

"Let's go." Lucy had been looking out the window and informing us of the deteriorating situation in the streets. Here in Denver many marched together towards the capitol. They thought to mob their leaders, as had been the case in many other countries.

The President was now public enemy number one.

Cleopatra walked to a tall cabinet on the other side of the apartment. She retrieved a shotgun and loaded a single charge. She grabbed a box of slugs and approached Charlie.

"Charlie, I know you were in the field, but that looks like a rifle to me, and you know I was always the better shot."

"But– but–"

"You know better than to tell her any different," Lucy interrupted.

Charlie reluctantly exchanged weapons. Lucy and I remained unarmed, though I was glad to have an escort. The four of us loaded the elevator, leaving the fourth floor.

The doors opened once, but a couple, with their young child, chose not to join our armed party. We passed through the complex doors and loaded into the Peterson-Chambers' van.

"Lucy, you drive. We'll take 70 East towards Strasburg, then 79 North."

"You're the boss," Lucy said as a grin curved her lips.

We exited the parking lot and began down the street with ease. If we could reach the interstate, I doubted we would have any trouble reaching our destination – a chapel in a cemetery. Cleopatra had downloaded a GPS app on her phone to better navigate the trails and find the new home of the monolith.

It was not long before we encountered a crowd and could hear sirens in the distance, as it became difficult for police to traverse the streets. Pedestrians did their best to hinder any authority figure – an anarchy incited to stretch and thin Earth's defenses.

The true invasion would begin soon.

A mob had gathered under an overpass, preventing the access we were so desperately searching for. Some were holding signs, many were chanting, and they were all angry.

"We bleed red! We bleed red!" I heard them shouting through the windows. I found it ironic that they would repeat the words I had used just days before to prove my humanity. They paid no attention to the van, and we were able to slowly progress through the crowd until we reached the center of the intersection.

Hidden from our view by the gathered crowd was an overturned police car – the source of the wailing siren. The body of a police officer was stood against it. They appeared to have been beaten to death and were heavily bruised. The blood on their face gleamed in the morning's light while it stained their uniform.

The citizens in this area chanted a variation of my call for unity:

"Pigs bleed blue! Pigs bleed blue!" they cried, not knowing the blood of their enemy had no color.

The car shook as those circling the destroyed vehicle turned to the Peterson-Chambers' family vehicle. They banged their fists forcefully, cracking the glass of the windows. They jerked at the door handles. Some wore masks, but most saw no reason to hide their face.

Lucy managed the turn and move past the upturned vehicle – the open trunk now in the air. The lights continued to strobe, adding to the anxiety of everyone in the van and encouraging the mob at the center of this mass hysteria. As we were forced to a complete stop, I saw the look of recognition I had been dreading as eyes on the opposite side of the window widened.

The man turned and silently informed those nearest him. They beat furiously against the glass together. It shattered, and they grabbed the hood of my jacket, pulling me through the new opening and constricting my neck.

I heard Lucy and Cleopatra calling after me as they unlocked the doors and attempted to open them, now finding the crowd pressing themselves to the doors.

"It's him! It's him! *The President!*" the first man called out to the crowd, rumbling in cheers. I felt as though I was about to be ripped apart by the many hands preparing to pass me back towards the gruesome scene we wished to leave behind.

CRACKOOM! a gunshot suddenly thundered through the excitement.

Some hands gripping me – pins piercing deeply through the skin and drawing blood – loosened as their owners turned to the source of the sound. Charles was sitting on the door of the van with the window down, reloading the shotgun.

"Let him go – and let us *pass*!" he ordered, aiming the shotgun towards the ground near the van. The area was cleared quickly by the pained ringing in the ears of those nearby.

"You'd shoot us, to protect one of *them*?"

"As is my *right*!" Charles shot back. "Black, white, green, or blue – he's still my Commander-in-Chief! And he may be the only one who can help us fight back!" The crowd did not seem swayed in this belief. He looked back at his comrade, nervously readying his weapon.

He gave her a nod, and Cleopatra made sure everyone got a good look at the alien technology.

"We captured this alien. We have his weapon. If anyone is saving the world today, it's *us*."

The man nearest her, holding me up, laughed.

"You can help us," she began, lowering the gun towards the heckler, "or not." He coughed once before he and the others returned me to my feet. The crowd parted, this additional confusion adding to their frustration.

I adjusted my ruffled attire and opened the door I had been pulled through. As I closed it, glass fell into the vehicle. Over the squeal of tires, the clink of spare change and broken bottles could be heard coming from the pocket of the door. The van shook with the revving of the engine.

The crowd parted quickly as Lucy did not intend to slow our acceleration. We climbed the on-ramp onto the interstate, comparatively open and desolate, and left the suffocating close quarters of my near lynching behind.

Not much was said during the drive, each of us reflecting on the revelations of the past few hours. They wrestled with their collective identity in the universe, now aware of life beyond Earth.

Nemi had shared in this struggle when we first saw beyond The Break.

Contrarily, I wrestled within myself to reclaim an identity stolen from me by my people. I had been forced to condemn humanity to an existence as a slave race.

I hoped I could earn my forgiveness before day's end.

CHAPTER XV

Reunion ALPHA

We left the interstate behind and took the highway. It was the middle of the day as we navigated a curve on past empty grain fields. We saw the rustic chapel – solid white and clean. There were no vehicles about nor any mourners that had found their way to the cemetery.

In the center of a dirt path, leading to the chapel doors, was the gleaming monolith.

"They probably *been* meeting here," Charlie insisted. "I'm sure this wasn't on their schedule."

Before any of us could protest our adopted plan, the doors to the chapel swung open. A man in camouflage fatigues and a tactical vest stepped out, though I could see several more bodies behind him.

They were armed, and some I recognized as members of the NIB.

I had another feeling of déjà vu, remembering a conversation I had in a dream while watching this scene from above. As I looked around us and saw several armed militant-looking types holding perimeter positions, confirmation swept through me as the feeling faded.

"Well, well, looks like someone brought us the President of the United States! How are we today, *Kriss Methius*?" they asked, their pale eyes piercing mine with a familiar intensity.

My companions and I grew increasingly anxious. I had informed them of the many Imprir operatives that had infiltrated every facet of their society – all of which knew my true identity.

"Why are you here?"

"I'm here to help these people!" I cried, nothing to wield but the reconditioning device. The man saw what I was clutching, and his demeanor lightened. "You would lead them to their *deaths*!" I finished.

"Lower your weapons!" As he turned his head, I noticed a scar behind his right ear. "You don't recognize your most trusted comrade, Uubiron?"

"Uubiron? I...you've changed."

"And you have not. Bold move."

He grabbed me by the shoulder and turned to look at my neck.

"I see...," he said, more to himself than I. "I guess *that* conversation will have to wait. Can't do anything about it here."

"Uubiron..."

"You look tired. Come – sit!" he instructed me, patting me on the back and directing me toward the chapel doors. "I've got quite a story to tell – you'll want to hear *all* of it." He stopped and faced me.

"It's all about *you*."

While I sat, he recounted his actions after the Vietnam incident. He apologized – repeatedly – to the point of annoyance. I assured him I held no grudge and reminded him we had little time. He looked at me silently, a softened look in his new eyes.

He agreed.

He truly believed he was helping me by revealing my intent to the Imprir, though his doubts began when forced to dispose of our followers, burning the jungle. The blaze consumed two more villages, and it was unknown how many more lives.

I had fought ferociously to keep my memories. After seeing me struggle with my conditioning for days, he begged the Imprir to return me home, using a substitute in my place. The Imprir instead continued to attempt several reconditioning procedures, my last episode manifesting in my refusal to take the life of Oscar Orwell.

He, too, was reconditioned and recommissioned for deployment at the close of the Nigerian Civil War. He took on the life of an oil field worker lost in an attack in Kwale – Adebowale Ikande.

It was this memory – forcing its way through as endless nightmares – that broke his conditioning.

After becoming the leader of Nigeria, he too grew fond of his citizens. He saw how a history of enslavement and oppression had broken a civilization much like the Old Government had ruled over Nemi.

He also saw their capacity to hope – what he considered humanity's greatest strength.

He explained that during Instruction, as our bodies were changed to appear human, a device was installed to assure our continued compliance with the wishes of the Imprir. It also served a secondary purpose – providing information to Militia Majora operatives off-planet.

He was forced to acquire help from the outside, operating through protected chat rooms made by those who believed what was now known to be the truth. They had been preparing for decades in expectation of a malevolent alien invasion – the many members of the MUFON network. They proved trustworthy and helped Uubiron to organize several storehouses worldwide.

"I've convinced some of the new class of Instructed – those coming from Nemi these days are much more *progressive*. We even have a member of the Moloke Institute to aid in their continuing metamorphosis."

He paused in his debriefing and looked across the headstones.

"Logaani...how did she die?"

I pointed at the weapon Cleopatra had placed against our transport.

"A Phalae posing as an Air Force officer – Yekk Agamo," I informed him.

His heart sang out a sorrowful song, the tune driving away even the crows.

"She is a good soldier," I added.

"They're always good soldiers! That's the problem! A soldier should know when to follow orders and when to follow what is *right*! But that is not the Militia way..."

"We are all born Militia."

"Not *them*," he stated, staring at the companions I had picked up. "But I've helped them build their own planetary Militia," he bragged as he raised a hand towards the chapel.

"Careful calling them that. If it has come to pass, it shall pass anew," I stated, using a Nemitian cliché.

"Yeah, history seems to repeat itself around here, too," he replied with the Earth counterpart. "After the High Inquisitor interrogated me *personally* – can you believe he left Nemi? – I knew Logaani and I had been discovered. When I had convinced her to join me, I removed the limiter behind her ear so she could move about undetected. She returned to Nemi and infiltrated the Temple." He paused to ensure his next point would imprint on me.

"Kriss, she learned much. What we know about this world – it's so little, we've been –"

"Sir, something's happening!" came a call from outside.

Everyone exited the church and followed the gaze of those assembled in the cemetery near the van and my party. Their mouths hung open as they viewed the scene above.

An irregular dark cloud had appeared above the fields across the highway. It moved slowly and with purpose. An even darker shadow could be seen within, a glint reflecting brightly off of a metal hull. The friction between the mass and the atmosphere ignited the exhaust fumes, leaving two large combustion units on either side of the cigar-shaped vessel.

Neon snakes streaked about the craft. As the cloud and bolts of pink lightning dissipated, a massive ship descended over the field. The wind began whipping around us, even at this great distance.

Charlie whistled a melody but was cut short as a blast exploded beside us and obliterated the monolith.

A TR-3BETA materialized above the graveyard. Several soldiers in Militia Majora fatigues leapt to the ground and readied their rad rifles. The forward weapon activated, a beam of plasma connecting the barrel and a body beside me. The light dissipated and left a blind spot in my vision, tracing its path.

I turned too slowly, only seeing a bright flash before Charles was no more. Lucy was diving in his direction, unprepared for the realities of our disparate levels of technology.

She swung her arms through the air her friend had once occupied.

"No!" Cleopatra screeched as Lucy rose to her feet, holding her arms out to her side. She looked over her shoulder at the party advancing from the tree line.

CHAPTER XVI

Lucy and Cleopatra Peterson-Chambers

"LUCY!" Cleopatra cried in a heavy wailing.

Lucy's eyes were wide open. She was raising her right hand to wave as the rad beam struck her. The bright flash shrouded her features in a momentary darkness. When it faded, in the infinitesimal moment before her body disintegrated, a single tear evaporated.

Cleopatra continued shrieking as they were locked in a vision of their lives together, the major acts performed by ghostly entities.

* * *

Arkansas, 2013

Cleopatra Chambers had recently been recruited into the Air Force. She was twenty-two years old, no particular career path ahead of her. From a small town in Arkansas with small town dreams, she wanted to work a stable job with benefits and there were few opportunities available.

She had saved some money after several years in the service industry, though it was not as lucrative as it once was. Customers still tipped, of course, but with the restrictions on restaurants, bars, and other service market employers during the pandemic, there were not enough hours *or* tables to support their workers.

Limiting seating capacity and enforcing curfews, the government seemed to do everything they could to starve her.

Factory jobs were the most stable until the CEOs outsourced cheaper labor after the minimum wage was raised. The corporations that remained began replacing every position with a mechanical arm that cost less in electricity than an annual salary.

A computer does not need insurance, bonuses, or other *superlative* expenses.

She had considered applying for scholarships to take classes but felt her trust was better placed in the United States military.

She was studying programming as her aptitude testing had directed. She had been given several technical manuals and often came to the local library for the seclusion. Public locations such as these were desolate, many readers preferring digital content or home delivery.

The loss of communal identity was not lost on her. Her nostalgia caused her soul pain, remembering the days of book fairs at school and regular trips made with her friends to the library. Taking nothing home, they would hide the books they were reading, bend the corners to save their place, and leave notes for future readers to discover.

This particular day, she recalled one of her favorite books – *Jurassic Park* by Michael Crichton. She had become a heavy reader as a child. As she learned much of these stories had been omitted, she found the joy of repeat readings of her favorite stories, ever finding something new.

She was looking to expand her awareness with a change of genre. She had seen many films after reading the book and was heavily disappointed with their lack of faith to the original storytelling.

Though she had enjoyed the movie adaptation of Crichton's *Timeline*, she had never read the book.

As she navigated the maze of aisles in search of her prize, she came upon the C section and ran her fingers along the bindings of several books before finding *The Andromeda Strain*. As she searched through the library's collection of Crichton novels, she found a space where *Timeline* should be. Resigned that the book would not be returned today, she exited the aisle and passed another youth reading at one of the many tables.

As Cleopatra passed, looking over her shoulder, she noticed the woman had stacked two copies of the book she was looking for next to a messenger bag. It was science-fiction themed, featuring a large logo Cleopatra recognized from television.

She stood there for several moments, secretly hating the stranger. They glanced up from their third copy of *Timeline*, and asked:

"Can I help you? I don't work here, the cart's just there."

Cleopatra looked at a cart filled with many titles stacked precariously. As she looked down a nearby aisle, she noticed an employee staring at the two of them and dropping a book before scurrying away.

Cleopatra turned back to the stranger, now looking up at her and smiling.

"Do you really need *all* of them?" she inquired aggressively.

"I suppose not." The woman closed the copy in her hands and slid it across the table towards Cleopatra. "I just joined a book club. This was going to be my suggestion. Have you read it?"

"No...I like Crichton."

"Well, you're welcome to join us, if you don't mind meeting at the shelter. I pick up the books here since no one can really buy them."

Cleopatra warmed to this personality. Most likely well-read – a sign of intelligence. She was charitable and chivalrous, certainly another attractive trait.

She blushed, considering the woman's appearance, quickly shaking her hair to cover her face and expression.

"My name's Lucy – Lucy Peterson. I hope you don't mind – we tend to take notes in the margins," she stated.

This addition guaranteed Cleopatra's attendance and soon their shared love of books would become a shared love.

* * *

A glow had appeared, wrapping over Lucy's shoulders then around her hips. The effects of the weapon were spreading, burning through her clothing and igniting the fabric. The hair on the back of her neck was singed, and a layer of thin smoke provided a hazy backdrop to her imminent death. Her skin was bursting in small bubbles as the tissue and blood beneath it reacted violently to the radioactive charge.

The bottom row of Cleopatra's eyelashes was soaked. She had dropped her weapon and was lifting her foot to run towards her doomed love.

Lucy appeared to be moving towards her, but this was the momentum provided by the strike. – Lucy was falling. I knew if I allowed Cleopatra to continue on, she would be there to catch the last of her wife's ashes. I feared for the survivor's safety as I leapt between them.

* * *

Arkansas, 2015

The millennial generation had been berated, psychoanalyzed, and dismissed, as any present generation is to be treated by the irrelevant.

February 6, 2015, Lucille Peterson and Cleopatra Chambers became Lucy and Cleopatra Peterson-Chambers.

They were the first in a long line at the county clerk's office that day. The fight was not over – there were many members of the district courts that could not separate their personal beliefs from their position to serve all citizens. The community faced much opposition.

Locally, a city council member had an outburst and attacked their server.

"Prove to me you're a woman!" the portly man cried.

They were glad to see him ridiculed by their local media.

Lucy and Cleopatra celebrated at a local bar and flaunted their new wedding bands. They played a game of pool over beers until two men approached them as though they were stalking some prey.

They commenced with the standard posturing and questions. When informed of the couple's recent nuptials, the presumed, "alpha," of the two promptly responded:

"Oh, yeah, I'm not into that *gay* shit."

They continued to harass the two women, oblivious of their hypocrisy and lack of tact – though a wedding band was hardly seen as a barrier to these boys. They followed the couple outside.

Cleopatra pressed against Lucy's ballerina frame, the dominant personality of the relationship. Lucy's hair, in a fashionable pixie cut, was enveloped in the afro her newly wedded wore at the time. They kissed, holding each other tightly.

The two boys watched lecherously and Cleopatra turned to them after their moment of passion to say:

"Now *fuck off*, or I'll kick both your asses." She had come from a more conservative family than Lucy, and has become one of the most proud women Lucy had ever met.

The two slender forms whimpered away, leaving these happy two alone to discuss their many plans for the future.

Their honeymoon would have to wait until after Cleopatra's return from Afghanistan – she left in the morning.

* * *

Lucy's hair was aflame. The black and boiling bits of skin clinging still to her face separated. The radiation wrapped around the skull and consumed abdomen.

She had separated in half, her melted torso bending in the wind.

Her limbs were glowing, and her clothes burned away. At each bent knuckle, the tendons had disappeared. Her fingers hung loosely, all but falling to the ground, elevated only by the passing of this moment in time.

Her face stretched horribly as she relayed the pain overtaking her. She was already dead, but it took time for this final signal to traverse her nervous system.

All this I saw by the light of the flames reflected in Cleopatra's eyes.

* * *

Arkansas, 2026

She knew Lucy had something planned as soon as they arrived at the bookstore. There was a grand opening today, but it had been delayed until two o'clock. No one was here at this early hour except the owner, who quickly rushed to unlock the front door. The woman did not speak and disappeared into the back of the store.

Cleopatra was having difficulty sleeping, and the previous night had been no different. As they walked through the store, Lucy led her to a small rack covered in two pastel clothes – one pink and one blue.

"I thought about what you said," Lucy began, referring to a debate between the married couple on their honeymoon.

They had decided they wanted to start a family. They discussed the benefits and moral obligations of adoption. They considered surrogate motherhood.

What they agreed was important to them was they raise *their* child – to what extent was available they wanted their genetic makeup to continue on, the great drive of life. The chosen method was in vitro fertilization. Most clinics were LGBTQ friendly though that had not always been the case.

They could research donors and compare their viability. Lucy was infertile – Cleopatra would carry the child.

That was not the only proposal – Lucy had been offered a job in Denver. The two of them could live outside of town to facilitate Cleopatra's finding work. Lucy did not mind the commute as she *loved* to drive and had recently reacquired the privilege.

They looked for a two-bedroom apartment two weeks later. Cleopatra had applied for a position as a process engineer. While the majority of her duties would be clerical, she would occasionally be required to oversee and maintain the new AI construction line.

This she considered humanity's last place in industry, her position soon to be completely obsolete.

She hoped that all her preparation and sacrifice for this job would be worth the transition. She had healthcare should the pandemic become more real for the couple. Her skill was highly sought after at this transitional moment in history and her benefits included a 401K – something she never imagined from her career.

Her employer was aware of her needs and promised to provide for them and more. She would retain her salary for six months after accepting the position – plenty of time to order furniture and books to read to the baby.

* * *

I grabbed Cleopatra and began turning her away from the scene. All that was left of her wife was a cloud of ash dissipating in the wind and a few chunks of burning flesh floating above the ground, yet to burn away. The compression of the many cells combusting around them usually ground the bone to dust, all that remained of the high-intensity reaction.

In Lucy's singular case, I witnessed her skull fall to the ground in this bed of ashes, a single crack appearing through the right eye.

I held up my hand and gave the Militia party surrender signal, reserved for practical training in competitive team combat scenarios – we never expected to be on opposite sides of the realities of our game. I held my hand over my nose and stroked an invisible crest.

The approaching party halted.

Cleopatra was in shambles. This morning she woke up with a sarcastic belief in extraterrestrial life. She enjoyed watching documentaries making connections between the magic of gods and our current technologies. She had always enjoyed the big science fiction blockbuster movies.

Now they were a malicious reality that had taken the only thing she had in this world – her love for Lucy, and her love for Cleopatra. Their only dream was to grow old together – their greatest hope to die hand in hand, peacefully.

Today, Cleopatra watched as Lucy suffered her last moments in anguish, nothing left to comfort the wife but her skull – ashes smoldering in the sockets of the eyes.

* * *

Colorado, 2027

It was four in the morning – If the baby was awake, mother was awake.

When Cleopatra hoisted herself off of the couch and called for Lucy, no one answered. Usually they made sure to go to bed together and talk about their dreams for the baby boy – they had just learned the gender the day before. Cleopatra was surprised to find the bed empty. She shrugged her shoulders and opened the fridge to choose tonight's meal of leftovers. While the mother had felt the first kicks only recently, the cravings were well established.

Leftover beef stew in the microwave for eighty-eight seconds was the favored menu item.

She turned on the television and sat down in their recliner. While Cleopatra was deployed in Afghanistan, Lucy had completed her education as a nurse. The residency program kept her away often, on-call for the many emergencies and extra classes she took.

Cleopatra quickly consumed the bowl, full to the edge and coloring the floor in a dark splatter. She promptly fell asleep, empty bowl in hand.

She was awakened by a light knocking on the door. She shook her head, rubbing her eyes with her palms to stimulate her senses. She walked over to the door and opened it to find the hallway empty. Cleopatra stuck her head out and Lucy jumped in front of her, scaring her.

She had been drinking. She and her fellow graduates had thrown themselves a party traveling from bar to bar. Cleopatra did not mind as much tonight, though Lucy had been going out often. She had to be sober and had hoped Lucy would join her – eventually. It was difficult for Cleopatra before the pregnancy, but she felt much healthier. She wished Lucy would see the benefit of a longer life *together*.

Lucy informed her she had found a position at a hospital in the capitol. They would need to move there in a few months after the position became available. She would replace a nurse practitioner. The job paid well, and Lucy could continue her residency program after Cleopatra returned to work.

Lucy was most excited about working within her passions and develop her specialty in caring for transgender patients.

Cleopatra had supported the two of them while Lucy worked around her class schedule. She returned from Afghanistan and worked overtime operating a press in the local manufacturing plant – the lifeblood of the community. When she informed her employer she could no longer work in such conditions because of the pregnancy, she was promoted to the position she had initially applied for – none were more familiar with the functions of the machinery yet to be replaced by automation.

Cleopatra did her best not to harbor any resentment towards her wife.

To prepare for the pregnancy, she had saved much for the in vitro treatments. Each was expensive and no more likely than the last. They waited for each ovulation period as the bills piled higher.

After several internships and lab jobs, Lucy could finally ease this burden. She considered this a sign to change many things, including her drinking habit.

She swore to Cleopatra she would be sober by the birthdate.

There was only one requirement for her acceptance of the position: they would *both* be required to get an inoculation. Medial staff were considered at high risk of exposure to the viral outbreak currently causing a quarantine order in high population areas.

Lucy promised it would be safe for the child.

* * *

I could only mourn the loss of this new friend as I watched their relationship end.

As I held Cleopatra up with my shoulder, a Militia soldier approached, several a short distance behind. He lifted his weapon, ignoring my surrender. We were lucky a shot came from the chapel.

"What are you doing out there?! Run!" cried my comrade Uubiron. When I was having trouble lifting Cleopatra up to carry her out of harm's way, he called after me, "Leave her!"

I ignored his comment and took two heavy breaths before tossing her weight higher onto my shoulder. I ran towards the small band of renegades behind the chapel doors.

* * *

Colorado, 2028

They had traveled to Texas, and they were excited about the pregnancy. To give Cleopatra a break from the long car ride, they stopped in Austin. Both from small towns, they enjoyed taking pictures and exploring the "big city." They stayed the night with Lucy's friends from college and slept soundly, comfortable in the large bed they were provided.

The next morning brought dread as they awoke to sheets glued to their bodies and a small pool of blood. Cleopatra was embarrassed. They quickly showered and in their anxiety left the sheets in the hamper. They reached home and were admitted to the emergency room.

The nurse that attended to Cleopatra's blood test had also confirmed the pregnancy and performed the ultrasound two weeks before. In that moment, Lucy leapt to Cleopatra's side, and they shared in their tears of joy. Cleopatra had gone through several cycles of in vitro treatment before she conceived.

They never knew how greatly they wanted to start a family until they believed they could.

The nurse returned and informed them they had lost the pregnancy.

Cleopatra wailed as every tear she had shed before paled compared to the torrent soaking her gown. Lucy did her best to hold back her own tear as the words, "I'm sorry," left the nurse's lips. Lucy leapt across the room to grasp her partner's hand and pull her into the safe space within her arms, sharing in her desperation. The fetus was young, but the pain was as great as having lost a child.

The loss was marked by the cries of the mother, pitiful and helpless on the hospital bed – her head thrown back and arms hanging loosely at her sides. The wife squeezed her right hand as the nurse attempted to comfort her.

These three had seen the whole of the soul's life.

They returned home and held each other in bed. Cleopatra cried for hours and Lucy joined her after attempting to console her beloved wife. They could not afford to miss any more work than they had for what was supposed to be a relaxing vacation.

The miscarriage had also come with a heavy price tag. Their public healthcare would have covered the cost of Cleopatra's doctor visits had the baby made it to term, including the birth itself.

It was decided if they wanted a child in the future they would adopt – Cleopatra could not undergo the pain of losing a life she carried within again. They had only researched their options when a conservative candidate was elected and his administration drew back on the available paths for the community.

Lucy returned to drinking, and tensions rose between the couple. The issue culminated when Lucy burst into their apartment early one morning.

Cleopatra was not alone.

Lucy insisted she was being unfaithful. Cleopatra assured her she was not, that this friend was only keeping her company after she woke up alone and unable to contact Lucy. He was a comrade from Afghanistan she had offered a place to stay – he had recently arrived at HOPE Campus where they continued to volunteer and provided books from the Denver Public Library.

If ever there was a legal case, Cleopatra would give instruction in using the available computers to navigate the infinite reference material available on government sites. While her comrade would not stay with the couple after many apologies, he knew a young mother and her child – a precious little girl named Felicia – needing aid.

CHAPTER XVII

Reunion BETA

"These people are gonna get you killed, Kriss!" Uubiron chastised.

"You used to call them animals."

His eyes held mine a moment more before he shook his head.

I pressed my back to the wall near his position at a window. The glass had shattered from a previous barrage and the opposite wall was covered in scorch marks from through-fire.

In the moments since Lucy's death, rain had reached the chapel walls – it pattered away a dirge. The water had accumulated as condensation against the large craft when it descended through the clouds. It was now sitting above the field adjacent to the cemetery, the rain carried by a localized gale.

Three Mirrit transport vessels modeled after the *Quest* and *Dark Eye* descended from shuttle bays that opened in the forward face of the mothership. Simultaneously, several of the smaller craft materialized above the trees. They fired heavy blasts that exploded outside the chapel walls. A direct hit to the spire opened a hole in the vaulted ceiling, a bright ray of sunlight shining through to better illuminate those fearful souls huddled around the podium.

Uubiron had truly followed in my footsteps. He recruited those that were desperate – with nowhere to go. I hoped that in the past thirty years he had better prepared them to defend themselves than I. On the floor, in front of the first pew, sat Cleopatra. She was trembling as I approached.

Several of Uubiron's followers had taken up small arms. The few AWOL NIB carried human weapons as well – Uubiron an automatic rifle.

It would not be enough.

I propped myself up and replaced the hood of my jacket. I approached the chapel entrance as my comrade called after me.

"Kriss! What are you doing?"

I looked back and nodded, one hand on the door. I shoved with all my might that the heavy wood would open wide. I dashed to the shotgun buried in Charlie's ashes. I reached it in two steps and leapt back into the chapel. I dodged several shots, the last connecting with my shoe, which I quickly kicked off of my foot before it disintegrated.

I would have preferred the railgun, but the vehicle was four steps further.

One of Uubiron's younger followers – human – stared wide-eyed at me. They knew of my alien identity and – as revered as Uubiron may be by his unintentional cult – he was *not* their President. I handed him the gun that he might survive.

The boy blinked once and trotted out the front door.

"Parks! Damn kid," Uubiron spat under his breath, hoisting his rifle and leveraging it in the window's corner. He fired wildly to provide a distraction for the brash actions of the boy.

He returned with one hand holding a hat on his head and the other holding the reassembled rail gun Charlie had recovered. The boy approached me and dropped it at my feet.

"You know how to use this, sir?"

"I do."

"Parks, you almost got yourself killed out there! What were you thinking?" Uubiron berated him.

"I figured, sir, that the most important thing is you two – the two of you can take care of us – we need you to stop them." He was rambling, a beat of sweat riding across the wrinkle of his brow before dropping and curving into his eye – he blinked again. I placed my hand on his shoulder.

"Young one, every one of you is important. That's why we've come to help," I assured him, looking back at my comrade.

The heavy fire had stopped. The three of us gathered at the window and saw that one of the transport vessels was on approach. The grass below was blown into several geometric patterns – true crop circles.

A hatch opened as the craft landed. I was surprised to see the Named Superior of the Militia Majora, Hi'Leng Khano.

Two more of our number had been lost outside of the chapel and those within its walls were terrified by their first vision of the Imprir.

She was short for a Nemitian, but she was stronger than *any* Militia soldier. Unable to see her features, one might think her form resembled that of an intelligent ape – perhaps the broad chested gorilla. As she removed her helmet, a rounded metal casing with two dark orbs to allow sight, a mist provided by her climate apparatus wisped away in the wind.

Her face was revealed to the humans.

Her purple scales were dull and flaky, a common disorder developed after radiation exposure during space travel. I am certain, however, this made her more terrifying to the gathered lesser species. Her eyes were moving independently of each other. Her thick red crest sat motionless on top of her skull.

The Named Superior stepped slowly onto the earth. Her battle-suit gleamed as she stepped out of the shadow of her drop ship, the *Vanguard*. Unlike most vessels which boasted the plain texture and dull colors of the materials composing their hull, this ship was painted a deep red, with gold stripes wrapping around the vessel and straddling the wings. The most defining feature was the image of a Korvinuundra, a Nemitian bird of prey similar to Earth's eagle, projected over the features of the ship.

At the head of each Militia Majora class were the "Children of Adunaan," as his generations of progeny would come to be known.

Each would kill the previous student-become-master. It was a role honored on our world by those that survived and those that did not.

The third Named Superior, Hi'Leng Khano, had adorned the stylized and ceremonial suit of armor for performing the rites of this station. It gleamed brightly in the sun, having never seen battle.

It was clear she did not expect today to alter this record.

"Kaancha Purseis, call for a restraint order and set those to stun! We don't want to waste potential property," she sneered. "And the High Inquisitor wants these *moonies* alive!" she added. She turned to the chapel. The wind died for a moment, allowing the creaking of the walls to carry over the silence. The Named Superiorr chuckled maniacally.

"Kaancha Uubiron! Kaancha Kriss! We have worked too hard and too long to change course now! You have strayed from the path! But I promise – we *will* set you right!"

"I'm sure she means to recondition us," I said to Uubiron. He shook his head, keeping his eye down his barrel and on target.

"I do not think so, comrade."

I looked out the window and saw the Named Superior waving for several prisoners to be lined up between themselves and the chapel. She took one of the few weapons between the four of them, a standard thirty-eight caliber revolver.

She readied the weapon behind one prisoner, who was begging for their life.

"No, please I --," was all they managed before a shot cut short their plea.

CRACK! The resolution of another life in terror.

* * *

I saw the boy, Oscar Orwell, lying on the stage. It was not a clean shot, but he would bleed out shortly – a painful death.

"Help...please," he muttered pitifully. He was blinded by blood loss and did not realize the one he begged for mercy from was his attacker.

Logaani Niih had lost that signature smirk. She glanced up at me, making the moment an eternity. When she looked away she raised the weapon.

* * *

"Please, God, no!"

CRACK!

I leapt out of the window, cutting my hand on the broken glass. Behind me, Parks gawked at the shimmering translucent blood left on the window frame.

I quickly bounded the distance between myself and the executions. I threw down Yekk's weapon before the Named Superior.

"I tried to surrender!"

"But you did not surrender to *me*," she replied, ignoring that she had yet to arrive at the battle.

I noticed none of our attackers had fallen.

"I accept your surrender – but what to do with you *now*, Kriss? I don't know that reconditioning is going to be enough." She approached me and lowered her face that I could feel her hot breath singing my senses.

"You *are* awake now, aren't you?"

"Can one ever know?" I replied. This caused the Imprir to chuckle.

"Not the response one would expect." Her face then became stern. "Does Uubiron agree with your surrender?"

I looked behind me to see my comrade step outside. He lowered his weapon and stood relaxed, shrugging as the humans had gathered at several windows to view Hi'Leng Khano's alien form.

"Good," she said, before waving to the two soldiers that had forced the prisoners to their knees. They turned their weapons on my comrade and his followers. I felt the sting as a wave of low-level radiation traversed my body.

I fought against the sudden weight of the world pinning me to the ground. I heard the Named Superior order everyone taken aboard the *Vanguard*.

"What about these two?" Purseis asked of his Superior.

"I suppose we don't need all of them."

CRACK!

CRACK!

CHAPTER XVIII

Solar warden

I could see nothing, though I was buffeted by maelstrom winds. I was blown backward, but no friction caught my form. I screamed, only to drown.

My feet suddenly felt a floor beneath them. Gravity arched my back, and I vomited lightning. The colors became a white spotlight, illuminating my bruised and battered back. I saw blood trickling from my nose on the reflective floor.

We could make it out of here, came a familiar voice. *But you don't know who we are.*

I turned and saw Uubiron caught in another spotlight, staring up at the source.

"What is happening?"

I fought the urge to squeeze my eyes shut and saw darkness smoldering from those that stared back at me.

The light only blinds us, burning through the void that which should have burned out long ago. Nothing, emptiness, *these are the absolute truths!*

Do you agree? the voice asked, now sounding off as my own.

"Yes!"

The darkness receded, and I shut my eyes tight as the light brightened to consume this space.

* * *

I awoke on the floor of a holding cell and my cheek took a moment to regain its form after detaching it from the cold metal floor. I knew I was aboard the mothership. My only company remained unconscious, though Uubiron appeared as though he had already been interrogated.

I looked through the barrier separating us from the detention hold of the craft. It produced a soft blue glow that illuminated the walls, faded, and appeared again.

Another holding cell.

I assumed our human comrades had also been abducted and hoped that Cleopatra had been among them. I mourned the loss of Lucy and Charles, swearing to any who would hear my pledge that I would protect her at the cost of my own life.

HACK! I heard behind me. I turned to see Uubiron pulling himself up, his back to the wall. He squeezed his eyes shut and showed his teeth, indicating some great pain.

"You're lucky you weren't the first awake."

"Comrade, I don't feel lucky. I'm still here," I replied. "Stuck with you, no less." A laugh pained Uubiron.

"I'm sorry to have betrayed you, Kriss."

"I can see how sorry you are, old friend. Did all know of my conditioning? Were *all* deceiving me?" I asked.

"Yes – and no. It took a long time before I saw the error I had made suggesting it – Kriss. I asked the Imprir to recondition you. I didn't want to see you *retired*." He could not look at me. "But I do not regret it, with the knowledge I have now."

"You were all *laughing* at me!"

"When you tried to leave, I wasn't the only one you came to," he reminded me. "Iniki – he couldn't *see* it then. But he would be with us now if he had just made it to the Sub-Base…"

"And Turrina – Chirrak?"

"Our little '*Resistance*,'" he said in his best impression of our French counterpart. "I could not have built it alone. But, Kriss, there is more you could not know, even now. And I fear you never will…"

"What is this new ambiguity? Uubiron, know it is unappreciated. I grow tired of vague statements and metaphor!"

"What did she say?" he insisted.

"Who?"

"The Named One! What did she say!"

"She...asked if I was awake," I recalled, though the stunning blast had made this recent memory hazy.

"So, she knows..."

"I *don't* know and that's what I told her!" I cried, growing frustrated. "Why this questioning Uubiron? Did they return you here to torment me?" He winced with another chuckle.

"Would they were able." He rubbed the back of his neck. "It doesn't matter much now," he resigned.

"Are we going home?"

I was searching for a silver lining to our capture.

"Home?" he said, staring at me in disbelief. "You really don't know, do you? Kriss, we *have* no home. Tell me – have you been dreaming?"

"*Shhh!*" I hissed.

A soldier was approaching, the echoes of their boots reverberating along the metal walls. The noises coming from the next cell ceased. I was surprised to see Yekk Agamo serving as our captor.

"Kaancha..."

"You can call me Kriss, Yekk Agamo," I replied, hoping this was the answer to my prayer. I had witnessed the compassion of this individual and knew of their disdain with the Imprir.

She wielded her rail gun once again. I grew anxious, realizing the heavy physical payload would reach a velocity capable of pressuring through the barrier.

"You returned this to me."

"Not purposefully," I admitted. "I had to steal it first."

"I'm taking it as a sign, *Supe,*" she replied. "I got a feeling I should help you."

"To what purpose?"

"I dunno myself. But..." She was reluctant to say more.

Her hand began moving before she was looking at it. The hand pressed the generator release and suddenly a breeze whiffed my hair. The pressure barrier stabilized between the cell and the rest of the ship.

"I think I might be dreaming now."

"Ah, the Alpha, Yekk," came Uubiron's voice from behind me.

"It's Kaancha, now," I added, winking at her, now blushing as red as her hair.

"Please, supes – you each tried to recruit me for what you said was, 'a higher calling. What did you mean?" she asked as she helped Uubiron to his feet.

"I'd love to tell you, but – might you know where they are keeping our things? We have a delivery to make."

We followed the soldier down the hall and passed the adjacent holding cell. Beyond the barrier, we saw several of Uubiron's recruits, all in light fatigues – missing were the NIB.

Cleopatra rocked back and forth, staring at the rough metal wall providing no reflection.

Yekk released the other prisoners and agreed to aid in our escape. A single guard remained with Cleopatra. As we walked, Uubiron and Yekk discussed their previous arrangement.

"What was claimed from the chapel?"

"There was no need for the human weapons but we stored them adjacent the plaza – we found one conditioning device. Turning to me, she added, "and the weapon schematics."

"What luck. We can break the encryption here and find a way off this ship."

"It will be a long trip...," Yekk informed us as we entered an empty plaza at the front of the ship. The paneling was replaced by thick glass, though blinds were currently raised, obscuring our view.

We approached a locked chamber and our savior entered a code into the pad before placing her thumb in the DNA analyzer. The doors grinded open, revealing several cargo containers within. Some were marked as explosive.

Yekk shuffled through the smaller units stacked on shelves along the wall until she found the correct ID tag. She opened the casing with a quick POP!

"This is what you're looking for?" She held up the conditioning device, causing Uubiron and I to duck.

She placed the device to the side. There was only one other item in the box – the encrypted flash drive.

"I'll take that," said Uubiron, swiping the drive from Yekk's hand.

He walked over to a terminal near the door. The purpose of these computers was to log inventory and communicate within the ship's network, though many guards downloaded a game or two. I was surprised to find they had USB ports.

"This is going to take some time," Uubiron called over the clack of keys, never turning back.

"We don't have any time! I was sent to collect you for *torture,*" she reminded us. "They'll be expecting us! The entire ship will be on alert if we don't show!"

"The interrogation room is on the other side of these cargo units."

"How do you know this?" I asked.

"I kept my eyes open. I hope you'll do the same – and you're going to have to stall. I won't lie to you Kriss," he admitted, a pause in his typing. "The Regent is here too."

The Regent was known for taking pleasure in extracting information.

"How long?"

"I don't know, but while you're in there, I'm going to push these crates against the back wall." He motioned to the explosive material.

"Won't miss that signal." I could read his plan without a word.

It was good to be with comrades, old and new.

"You two go," Uubiron insisted. "Yekk, hurry back and help me get ready." He reached out to her, and she blinked once before they clasped arms. "Welcome to *le resistance.*"

As Yekk and I bounded the doorframe, he called after me.

"Kriss, stay strong. Know there is much more on the other side of today."

I nodded my recognition and followed Yekk to what would become my torture chamber.

"Before we go, there's something I want to say."

"Permission granted."

"I didn't want those men to die. We should have had our blasters set on stun to begin with! That's Militia protocol on Nemi – and that woman...she's been like an invalid since she woke up," she informed me.

I swallowed the guilt attempting to drown me. Yekk paused at the hollow sound. I watched as she entered her code and the doors slid open. She led me into a large open space.

There was a light on the floor where a raised platform stood. Yekk gripped my arm tightly, committing to the performance, and led me to the central stage. As she exited, the surrounding air flashed blue – I was trapped within a force field.

Two forms were then illuminated in the darkness.

Sitting in raised thrones, I recognized first the Named Superior with whom I had recently faced on Earth. Her companion, Ugden Mirrit, sat smirking above me. They both removed their breathing apparatuses, instead sitting them on the long table in front of them.

I realized the ship was designed for human transport and operation.

"*Ohh*...the expressions they can make!" chortled the Regent. "Let us begin!" he insisted, holding a hand above the paneling before him.

Large by any standard, his features were distinctly hideous. Swelling under his scaly skin caused it to split in many places. One such scar bounded his fat crest, the top too heavy to hold itself up and resting on his two horns, one piercing the pustual. He boasted two rings on his lower lip, the weight revealing jagged yellowed teeth.

"Come now, Ugden, we have one more arrival incoming," the Named Superior insisted as a tall glass chamber rose from the floor before me. It was several inches wider and taller than my form – just large enough to hold one Nemitian.

A second pillar of light appeared, blinding me momentarily. It began rotating within the chamber, continuing the assault on my vision. As its speed increased, it appeared to fill the chamber completely. A loud whining reverberated off the walls of the room before the chamber disappeared in a cloud of vapors. A shadow moved towards me, illuminated as it stepped onto the platform.

Through eyes still burning and tears staining my cheeks, I saw Hi'Ran Akeht, the High Inquisitor of the Imprir.

"If only I could truly be with you now, *Kriss Methius*." He held an open palm before me. "This is a projection – I have duties to attend to here at home. But," he continued, the projection sweeping the arm downward and striking my cheek, "it is fully functional."

I prayed Uubiron would be quick.

"Kriss Methius!" the High Inquisitor called to the other Imprir present as he stepped off the stage. "You have always been a trouble-maker!"

Lights surrounding the room provided dim illumination, and I saw that the base of the glass chamber the apparition first appeared in was moving with it. This must be the source of the hologram. Wherever the High Inquisitor was, several sensors followed him as he strutted around an empty room somewhere in the city of Ammagadan – the capitol of Nemi.

"You will *not* disrupt our plans for Earth!" came the Regent's guttural bellows. "Nemi requires the resources of this world *now*, and you have revealed us to the Loarkeen!"

"Our infiltration has become *null*!" cried the Named Superior. "You think to save humanity by inciting a war they cannot win?"

"Living as a slave is no life at all!" I cried back. "Believe me, they would rather *die* –"

Suddenly, the air before my lips and around my jaw shimmered. A band appeared and tightly bound my objections. I looked down to see the High Inquisitor had raised a hand, this projection able to silence me.

"Look how he speaks of life! The Militia was created to serve – *you* were created to serve! You have been given a purpose! Why would you abandon these convictions?" I made no protest, showed no struggle, though I could hardly breathe.

The High Inquisitor waved his hand, and the binding dissipated.

"Speak!" the Named Superior ordered from her high chair after I remained silent.

I looked to her, then the Regent, before returning my gaze to the High Inquisitor before me – it was to this one I would argue the morality of their venture.

"They were never my convictions – *my* beliefs! I had been conditioned against my will! How is *any* Militia soldier to believe the will of their Superior to be true with this history of subjugating your insubordinate? If my standards were too high – if I would not follow orders – then why use me for this? Have you buried my memories of such crimes that I must suffer so?"

"What information do you require of him?" the High Inquisitor asked the hulking form of the Regent in its high chair.

"What do you know of our intended operations?" the Regent requested of me. "We are aware tyou have been in communication with Kaancha Uubiron and his agent, Logaani Niih."

I refused to respond, staring at the floor and mourning the loss of my comrade. The form of the High Inquisitor approached and slapped me. I fell to one knee. I kept myself off the ground with a fist.

"Answer him!" Whatever I thought I had seen in his holographic eyes was gone, replaced by a sinister glee.

"I know that our plan to eradicate the humans changed after the Temple was consumed by the Imprir. Until recently I...had *forgotten*," I continued, causing a smirk to curve the edge of the High Inquisitor's lips. "Our new orders were to infiltrate the world's governments and industry to hoard wealth, forcing the peoples into slavery. We were to remain undetected." Their silence begged I continue.

"The Moon Base was constructed to observe their world prior to invasion. A geological survey revealed elements in high demand and nearly depleted on the home world. The base was converted for mining and ore processing, though it was soon abandoned for another location."

They continued to observe me silently, informing me that there was more they believed to have been leaked by Uubiron.

"I know nothing more!" I cried truthfully. "I began Instruction and my focus was the policy and industry of *this* world! Any knowledge I had was buried under my conditioning!"

The gathered Imprir laughed in a horrid chorus. The elation caused the Regent to cough before he could speak again, spittle dribbling down his wrinkled beard of skin.

"Yes, yes, the process often leaves a memory here or there lost *forever.*"

"Yourself and Uubiron were the missing pieces that we may begin the invasion. We can reclaim control over the nuclear arsenals of Earth and guarantee their fealty!" cried the Named Superior, slamming a gloved fist onto the table in front of her.

The High Inquisitor remained silent, his arms crossed behind his back. While the other two spoke, our eyes were locked in a contest of wills.

"This one has spent enough time among the flock. Let us return them to their chosen comrades believing they are once more one of *them,*" I heard the Regent suggest. The High Inquisitor moved towards me as my eyes succumbed to their weight.

"We know the weapons manuals remain encrypted. The Loarkeen cannot defend themselves, as Uubiron's band of degenerates could not defend *your*selves. This ship was built with them in mind, you know," the High Inquisitor confirmed. "We needed to bring them – *here.*" He placed one hand on my shoulder and the other raising my chin.

The ceiling opened and revealed a curved layer of thick glass. Beyond the windows was the largest planet I had ever seen. I knew the autumn streaked gas giant to be Jupiter, with a view of its famous Red Spot.

On Nemi, there was *always* a large planet swirling with colorful highways of gas in the sky. I had lived thirty years at my Earthly station, however, and the view left me in awe. I imagined this was the feeling of those first Nemitians to navigate these worlds – the crew of the *Dark Eye.*

I shuddered as I felt the movements of the hologram behind me like a dark wind raising the hair on my neck.

"We'd like to thank the two of you for providing us with the first lot." He squeezed my shoulders tightly.

"Shall we *continue* this interrogation, Inquisitor?" insisted the Named Superior.

"Yes, our schedule *has* been forced, thanks to you. I shall leave the rest in your capable hands," he added before disappearing in a flash of light.

"May I?" requested the Regent.

"Have your fun."

The Named Superior got out of her seat and reattached her breathing apparatus. She began descending from her raised position. The Regent giggled as though a youngling with a new toy. He palmed a panel integrated into the table.

A shock wave of radiation sent burning sensations throughout my body. I crumpled immediately to the floor. The Named Superior stopped her exit to share in the Regent's enjoyment, laughing at my cries.

The Regent released me from my torment only in reaction to the sound of the wailing alarms. Yekk reentered the room.

"Imprir! Uubiron has escaped, releasing the human prisoners!" she informed us all. She gave me a nod, imperceptible to eyes not searching for the sign. I remained on the floor, acting as though still pained.

"Where are they now?" the Named Superior cried, drawing her weapon.

"Right here," answered a shadowed formed behind the opening doors nearest the Imprir.

She shouted a battle cry and lunged towards my comrade.

Uubiron stood at the wall panel opposite the doorway, quickly keying a command with one hand while holding his recovered rifle in the other.

The room shook as an explosion behind the raised platform vaporized the wall. The blast obliterated the floor, and the Regent disappeared into the cloud of dust.

His cursing and denial of this end was cut short by the sound of a boulder splashing in the mud – the Regent was no more and the Named Superior was buried under more of the rubble.

The containment field failed.

Both Uubiron and Yekk rushed over to catch my limp form and return me to my feet. They scrutinized the damage they had caused and determined the ship's capabilities had not been hindered. The control room was on the opposite side of several hallways and other rooms, though the blast will have been heard throughout.

"I'll release the others. We can take this ship!" pressed Uubiron.

"No casualties, Comrade," I implored him. "We are all victims of the Militia." This gave him pause.

"Believe me, I know. Then we can *try* to take this ship – *Supe*."

"I want to take the woman – Cleopatra. I want to return her home."

"How?" asked Uubiron.

"I can fly, Supes," Yekk interjected, raising a hand. "We can take your friend and any injured on transport."

"Good."

We headed back to the holding cells and gathered the troops. Cleopatra was more aware and agreed to come with us, helping one of the two injured in our party while I assisted the other.

Were returned to the cargo chamber. In the blast, the locks for the pleated metal blinds had failed as well. The large windows of the plaza revealed to its passengers how far they were from home. They had to be herded along into the storage room holding their weapons.

Their enemy would have superior firepower, but I had faith in Uubiron's tactics.

"Before we part ways, know that we have been deceived from the very beginning."

"Why has your deliberation been so cryptic, Uubiron?" I begged him.

"It only seems that way to you *now*," he insisted. "Just...please, should we both survive this journey, I can lead you to the truth."

"You sound like a Seeker."

He chuckled but would say no more.

"We must hurry," came Yekk's voice in the doorway. Cleopatra had found a first aid kit and was bandaging the wounded.

We exited the hold, Uubiron and I clasping arms before separating. Shouts were heard coming from the interrogation room through the newly made hole in the wall. Yekk, Cleopatra, myself, and the injured headed in the opposite direction.

We met no resistance navigating to one of the three hangars I had observed when the ship had descended above Bennett Cemetery. There was but a single engineer present, taken out by Yekk in a stealth maneuver, as we had no rad weapons to stun them.

The hangar held two TR-3BETA stealth craft and the *Vanguard*. We boarded the larger Mirrit vessel and Yekk prepared for liftoff. The onboard computer allowed her to access the controls, and the doors opened.

Those gathered were startled by a loud banging from outside the ship. A view screen was activated, revealing the Named Superior holding a rad rifle and slamming it against the doors of the craft. What remained of the glorious ceremonial armor of her station clung in shattered pieces to her skin, fused with scale in several places.

She backed away and attempted to shoot the outside paneling. Yekk deactivated the locks to prevent a power surge. The Imprir member fired repeatedly at the hull, but in vain. The rifle was heavy in her arms and her breathing was labored. The apparatus created to assist Nemitian physiology was missing, surely destroyed in the blast thought to have killed her.

The ship's antigravity systems were charged and activated themselves, raising the ship off the floor of the hangar. Slowly, we moved towards the blue barrier, retaining the atmosphere within.

We passed through as if there was no barrier at all, the Named Superior screaming silently after us into the void of space – we were free.

Yekk accessed the navigations system and set a heading for Earth. While the mothership had reached their present position orbiting

Jupiter in a matter of hours, the trip would be much longer on the transport vessel.

We would make a straight shot, constantly adjusting for the motion of their destination, as there was no time to pack extra rations. They were lucky the electromagnets of the CFG only required the electric charge from solar batteries.

"We may be able to use the planet's gravity to speed things up," Yekk suggested.

"Whatever you can do."

The ship oriented itself to face the sun as we began our journey. Cleopatra helped the injured followers get comfortable in the back of the ship and joined the two of us at the control paneling.

"Cleopatra, I – I don't know what to say."

"I want you to say you're going to *fix* this. You are no longer just responsible for the lives of me – of my *Lucy*. Not just the American people, either. I'm going to make sure you do," she assured me while stabbing at my heart through my chest with her finger. She searched her pockets, finding that the Militia had seen no reason to remove her wallet nor her pack of cigarettes.

She removed a photo I could not see.

"Cleopatra, please."

"Don't, '*Cleopatra pLeAsE,*' me, *Chris*," she snapped, hiding the picture. She was shaking. She was nervously sparking the flint of a lighter she kept in the pack.

"These materials are highly reactive – explosively so." I motioned to the wall compartments containing the elements necessary for the ship's function.

She grunted, placing the lighter and pack of cigarettes in her pocket.

"You know, –," she began.

A high pitched beeping interrupted my now companion, incessant in its announcement.

"You *know*, Mr. Alien–," she repeated.

"What is that?" I asked our pilot, my hands covering my ears. Yekk adjusted the volume to ensure she was heard.

"It's another ship, Supe," she read from the sensors. "The Superior must have followed us!"

CHAPTER XIX

Europa

The pursuing craft – a TR-3BETA – was much swifter than our transport vessel. While the CFG engines were the same on all craft, the lesser mass of the TR-3BETA gave less resistance to the engines and allowed for better maneuverability.

"Can you at least keep us straight?"

"I have piloted one of these before...I think," I answered.

This raised no great confidence in my companions.

"It'll have to do." Yekk accessed more of the controls and I heard the cargo bay door *CLANG!* as they began to lower.

"What are you *doing*? We have injured in there!" Cleopatra cried. "I may not know much about all this, but I know they'll get sucked out! That's why it's called a vacuum!"

"I activated the force field. They will be fine. Can you keep us *straight*!" Yekk repeated.

She left the pilot seat, and I sat in her place. I reached out for the hand controls, the ship jerking as I grabbed them simultaneously. I slipped my feet into the strapped pedals. Yekk showed me the gauges for speed and acceleration, then grabbed her weapon from beside me.

"What are you going to do with that?" Cleopatra asked.

"It's a railgun – similar to your Earth projectile weapons," she explained while loading a wide clip. "The charges will be fired at a high enough speed. The pressure against the force field will be too great to stop it from passing through."

She continued priming the weapon and checking its functionality. She walked to the door of the cabin and used the paneling to open it, turning back to us before attempting to shoot at the pursuant craft.

"No wind out there, no localized gravity – nothing to affect my trajectory. Perfect conditions."

The two injured humans had frightened looks on their faces as they stared out into space, backing away from the crackling barrier. The door was fully lowered, revealing the ship gaining on our position.

Seeing it with our own eyes, it seemed much closer than it appeared on sensors.

Yekk braced the weapon in a corner of the door frame and her body against the wall. I saw her chest expand as she took in a deep breath.

A faint red light appeared on one side of the ship behind us, then the other. These lights erupted in size and brightness. I imagined the blasts screaming through space as they approached their target, knowing they silently hounded the rear of the ship.

"Fuck," I heard Yekk and Cleopatra say in unison.

Yekk squeezed the trigger, her weapon emitting a buzzing that rapidly grew in volume before a charge lit the length of the barrel blue and fired its heavy charge. A blast confirmed her shot connected with one missile. She loaded a second charge and took aim at the second javelin.

Her weapon hummed with the tension we all carried in our spines and the missile blasted the hull of the ship.

While the rad charge would not pass through the field, it disrupted it momentarily. The vacuum of space acted quickly enough to inhale the two passengers in the hold and causing Yekk to grip the door frame. She slammed a fist against the panel to close the door.

The engines had been damaged, and we were moving solely on momentum. It was unclear whether Yekk had crippled our pursuer, but with no defenses of our own, we would not clear the planet.

"Activate the fuel systems," Yekk ordered. As I sat with my hand in the air searching the paneling before me confusedly, she grabbed my shoulder, crying, "Out of my seat!" I was glad to be rid of the responsibility and did not mind.

"These transport ships still have jet propulsion back-up systems should the CFG fail."

Yekk thrust the handles forward, increasing the fuel combustion. My legs shook, reverberating through the rest of my body as I crawled to a side seat and strapped myself in. I could still hear the beeping indicating the following craft had not shied away.

Yekk pulled back on one handle and the ship fought the gravity of the planet to pull away – classic slingshot maneuver. This rocked the craft, and I had to assist Cleopatra in securing herself when she reached the seat next to me.

The quaking ended. We passed over several asteroids and other debris. We could see a moon slowly growing in size out of the forward view port.

"Europa. They were going to bring humans here and mine the planet's atmosphere. It is a much more rich resource of *Helion*, and the water ice of the moon can be used to power the base."

"Yes...they could build a base deep into the ice, melting it and using the currents below the surface to power hydroelectric generators. I remember the Imprir considering this plan, but they found there was no way to staff it."

"Well, they figured it out, Supe."

I now realized why our plans for humanity had changed. The mining of Helion– its vast applications ranging from radiation containment to focusing energy from nuclear power sources – was the driving force behind all industry on Nemi – all products were made to use Helion batteries, patented by MoloCorp. Nanotechnologies were the most lucrative business, Helion in the form of moon dust used in its development.

The greed of the Imprir knew no bounds even here, beyond the Break.

"We can hide there – its uninhabited now."

"If she persists, we outnumber her, two-to-one."

"*Three*," asserted Cleopatra. "This is the one who killed my Lucy, right?"

"Yes."

"Well, I'm not staying on this ship." She was still shaking, but her eyes showed she no longer felt as though she was a burden.

She was determined to claim this vengeance.

We approached the dark moon. It did not have an atmosphere, though there was a single cloud covering several miles of the surface. It was this cloud we descended through.

"The platform is active?"

"Yes – and abandoned, I assure you."

"It cost too much to mass produce the breathing apparatuses. That's what gave them the idea to find a new workforce."

"You mean us – *humans*," Cleopatra said.

"Yea – sorry."

"Don't worry about it. I have to believe you're like us – that there are more like *you*, who see us as people, and not *tools* to be used at your leisure!"

The cabin grew silent as we cleared the cloud cover. Below us was an enormous mechanical structure bounding a gaping pit that bored into the water ice, forming a thick layer of Europa's crust. The edges of the facility emitted exhaust vapors rising from the core of the moon. This was contained and recycled by an atmospheric force field, powered by hydroelectric generators.

We entered the chasm, and Yekk directed the ship to a landing pad extending from the wall. We landed and exited the craft. We entered a checkpoint station that provided security for the active base. Several cargo containers of various sizes were littered along the wall near a desk.

We could see another shape descending from the cloud – it was moving erratically.

Yekk dashed to a panel and attempted to use her access code, but the screen flashed red several times before remaining so.

"Shit!" she called out, slamming a gloved palm against the wall. "She must have signaled the base's computer to lock us out!" We turned to the window and saw another pair of missiles fired from the craft.

"Take cover!"

We dove behind the terminal as a blast shook the structure, striking the *Vanguard* on the landing pad. Heavy bits of ship crashed through a wide window, atmosphere whipping through the security office. A wing snapped off with a high-pitched grinding that shook the teeth. It struck the far wall, blasting a hole through it.

"Come on!" Yekk cried out over the whipping winds. I grabbed Cleopatra by the arm and we entered the hallway as several blasts continued to rock the facility.

Yekk led us to another such checkpoint, the doors requiring no identification to exit the facility. From this position we saw the Named Superior's craft wobbling as it made a rough landing on what remained of the pad.

Behind us, Yekk and I heard soft metal grinding and the hiss of gas – we turned to see a shaking Cleopatra flicking her lighter.

Yekk's eyes widened, and she looked around for any of the containers we had seen in the previous office. She rushed over to identify their contents. She grabbed a small case by the handle and dragged it over to us.

"Let me see that!" She held out a hand and motioned that Cleopatra hand her the lighter.

"OK...," Cleopatra replied, placing the lighter in Yekk's palm.

Yekk opened the case and removed a vial of a dark liquid.

"This is pactwi oil – to grease the gears."

I had become quite familiar with the fuel used to enhance human weapons.

She opened the vial and poured some of the liquid on the floor. She activated the flame of the lighter and ignited the oil in a bright green flash, a crack appearing on the floor.

"If I douse one of these charges, and hit one of the larger casings I can –," she began, interrupted by the sound of feedback as the station's general comms were activated. The Named Superior wished to be heard in every room of the base.

"You have shown great resourcefulness in your choice of friends." We could hear the rasping of her breath, injuries accrued from our escaping the mothership.

"*You*, Kriss Methius, have been a fine soldier. You have served well, assisting in our deception of Earth. You would have been welcomed home as a hero, confirming our legacy! But you *chose* instead to defend the enemy!" she cried. An alarm began wailing and the dim lighting was replaced by the standard emergency red.

"Now you can die with your pets!" she cried, the comms now taken over by a robotic voice:

SELF DESTRUCT SEQUENCE INITIATED
FIFTEEN MINUTES TO EVACUATION

"We have to get back to that ship!" I insisted.

"The Superior will be ready for us," said Yekk, who had been dousing several charges in the pactwi oil.

Cleopatra, having remained silent in the excitement, strode over to our companion and grabbed her weapon. She loaded a single charge as she stepped outside onto the landing pad.

She was still shaking.

The weapon hung loosely in her hands as she stared toward her target. She loaded the charge and took a single drag of her cigarette. She placed the burning tip to the charge and ignited it before sealing the chamber. The cigarette she tossed over the edge.

She became still.

She raised the weapon, placing the scope just millimeters from her eye.

She squeezed the trigger, causing the rifle's mechanism to flash blue as it charged and fired a bright green bolt. Moments later, a second explosion rocked the station. Yekk and I rushed to the window and saw the second ship falling into the chasm.

"Fuck!" the shooter cursed.

Yekk kicked another charge, which glided to Cleopatra's feet, leaving a trail of thick oil. She loaded and loosed a second shot, then another.

No contact was made by either volley.

"Fuck you, goddamit!" she screamed as tears filled a puddle at her feet.

She tossed the rifle aside.

She was trembling with rage – not fear.

"Nice shot."

"I missed," she cried. "I've failed her again!"

"We no longer have a way off of this station before the self-destruct sequence is complete," I informed Yekk as I retrieved her weapon

As if to confirm my statement, the automated voice returned:

SELF DESTRUCT SEQUENCE INITIATED

TEN MINUTES TO EVACUATION

"Stay with her," I ordered before rushing down the hall.

I peeked around the corner and heard the rustling of debris.

HACK! came the Named Superior's unsteady breaths. I considered her weakened state my advantage as a human would be no match for a trained Nemitian.

She was holding herself up on a fallen boulder of the ceiling. I crept around the edge of the room until I faced the rear of her armor. A crescent blade glinted under the plate in the flashing red lights.

I reached out for the dagger and held my breath, tensing every muscle for a final quick swoop to end her life.

The armor burst forth, the Named Superior braced with her back against the boulder. I was launched several meters with the heavy bits of armor clattering around me. She drew the dagger and charged at me, bellowing a battle cry like a dozen horns erupting from within her crest. Flakes of scales loosed themselves from the irradiated skin.

I backed myself into more of the fuel crates and dodged a high blow. The ancient weapon of forgotten alloy punctured a crate, the pactwi oil drenching both myself and the Named Superior.

"Why Ahket allowed you to *continue*," she screeched as a swipe covered the floor and tore at my ribs. My blood mixed with the oil.

"This experiment is over!" her last call before a lunge forward holding the blade with both hands.

I rolled past the berserker and shoved her aside with my weight behind a shoulder. The slick floor threw her down as I crawled away.

There was equipment loose on the floor and I found a mining laser within my reach. As I squeezed the handle, a flickering blue neon projected itself several centimeters. I remained crouched until the Named Superior's shrieking deafened my own thoughts.

I twisted myself onto my back and bored a hole through the enemy's stomach.

The pactwi oil ignited a bright green that snaked across the scales of the Hi'Leng Khano. Her bloodlust became confusion. She backed away, swatting at the flames. When they survived her assault, she stood staring at her burning hands. Her crest erupted and shock gave way to pain.

With a final screech, the Named Superior of the Militia Majora collapsed in an emerald blaze.

I turned to see my comrades arriving. The door that would not allow us passage earlier opened and closed indiscriminately. As they searched the room, they found our battleground. I nodded toward the body.

We were relieved to find the security terminal intact, though access had been locked.

"Hope is not lost," I informed my confused companions, raising the instrument I had found before.

"A mining laser?" she inquired.

SELF DESTRUCT SEQUENCE INITIATED

TWO MINUTES TO EVACUATION

My companions made no comment as they awaited the completion of my grim work.

The fire had burned out, but I did not find the body of a large Nemitian form, nor was the Named Commander's armor present anywhere in the room. The remains, clearly an organic mass, were moist in oils and *alive* – pulsating a wicked heartbeat.

I lifted the mass and carried it to the control station.

SELF DESTRUCT SEQUENCE INITIATED

ONE MINUTE TO EVACUATION

I approached the terminal, suppressing a facial tick I had only just developed as I placed the grotesque mass on the DNA analyzer. When the screen's back light was again blue, I backed away as Yekk rushed to deactivate the sequence. Her eyes remained wide as she fought to keep them forward.

SELF DESTRUCT SEQUE – KZZT

She succeeded, falling back on her heels and seating herself with her arms supporting her, exhausted.

"We should have control of comms too," she said. "We can signal the mothership to see if Uubiron and his party have gained control."

I stepped up to the paneling and attempted to contact our comrade. As I opened the communications to receive, I saw that there was an incoming signal. I opened the channel for all to hear.

"—picked up some trouble. Are you in need of assistance?" came Uubiron's voice.

CHAPTER XX

The Break

When we arrived in the plaza of the Solar Warden, Uubiron was explaining the workings of a rad rifle to one of his human followers.

"Since we built these this side of the Break, it's a *radeon* source refracted through the dust –," he began. "Comrade!" he called, leaving the young woman when he saw my approach.

Uubiron and his followers had taken control of the vessel. The Militia soldiers aboard quickly surrendered at the discovery of Ugden Mirrit's death and their abandonment by the Named Superior.

"Any casualties?" I asked.

"There are always casualties, Kriss," he replied. "Our attempt to lessen them only cost us lives, so we returned in kind," he informed me, though my expression remained that of disappointment. "This is *war*, Kriss!" he reminded me. "Don't you forget that!"

"*Life* is war!" I said. "But what's done is done. You have decrypted the schematics?" I inquired further.

"More than that," Uubiron responded, spreading his arms as he took three steps away from me. "This ship was built to enslave the humans – now it will serve as their salvation!" he cried confidently.

I could not describe the feeling swelling within me as I watched the gathered Nemitians and freed human prisoners cheering together.

"Then we return to fight!" I cheered as well.

"No, my friend," Uubiron began. "Your journey is just beginning."

Uubiron retrieved the Named Superior's remains. Lacking the disgust of those around the room, he tinkered with something below the fleshy mass. The tissues and oily substance retracted. What remained was a small cracked crystalline sphere.

A common sprite companion.

"What am I seeing?"

"A soul, technically," he replied. "Specifically, the soul of Hi'Leng Khano. The Imprir don't want only Earth, Kriss – they want the *realm*. This is the only way we can exist here as Nemitians."

The new purpose of the Militia Majora was revealed to me.

"This can be programmed to appear human as well?" I deduced. Uubiron nodded and handed me the damaged technology.

"You must take this to the High Inquisitor. Only they can truly open your eyes," he insisted.

"Cryptic, still, at the end?" I inquired.

"Apologies, but it would be impossible to explain," he said, reaching for his backside pocket.

His hand reappeared as if he was holding an item between his palm with his thumb.

"What do you see?" he asked me.

"Nothing," I replied honestly.

"As you would," he said. He then grabbed a cloth from a pocket on his tactical vest, reclaimed along with confiscated weapons. He wrapped the cloth around the invisible item, giving it the form of a LightPad.

"You must return through the Break with this," he began his instructions. "Seek the First, only they can free you as they have me. You will also need this," he added, returning the conditioning device Logaani Niih had used to restore my identity.

"Kriss, there is a reason we are averse to the idea of enslaving the humans – we have lived this tale before," he explained. "The Imprir is the enemy to humanity *and* our people."

"Our service in the Militias –," I began.

"No, Kriss! This goes *beyond* that. Each territory had its own culture to protect – its own narratives of history. When the Imprir was formed, they *agreed* on a common history for our world. They enlisted the Speakers of the Temple to ensure it was believable – analogous to The Word!"

"This is known even to younglings!" I protested, offended at my perceived naivete.

"The truth cannot be debated, Kriss!" he cried back. "Before facts could be logged and events studied through firsthand accounts, our history was taught through song and folklore – much like on Earth. But The Word was known to be fables, meant to teach the true values of life – compassion, curiosity, self-sacrifice!"

"Knowledge is not a corruption of the soul!" I shot back. "The Imprir have many faults, but the removal of the Temple's power is *not* one of them!"

"Listen to yourself! You know your enemy, yet you defend their policy! Do you not wonder why?" he asked. "Do you not wonder to what depth their control has burrowed?"

I did not.

I wished this journey through my memories would end. My identity had been revealed to me – I could return to Earth with my comrades and help the humans build a defense network. I could lead the rebellion and become a *true* hero.

I knew these to be selfish desires I must overcome. Uubiron would perform the role efficiently, his leadership evidenced by his growing followers – human and Nemitian. With our defection, the Imprir had lost control over Earth's present nuclear arsenal. Yekk had also pledged to assist in the planet's liberation and provide intelligence on the Militia Majora's current movements and targets.

"Where are the NIB that you had recruited before our capture?" I inquired.

Uubiron's eyes fell to the ground.

"They were reconditioned after our arrival on the Solar Warden," he informed me. "They could not be convinced and were led against us by *this* one," he spat.

Purseis Comaan sat bound and kneeling beside the bodies of Nemitians at several stages of the process of their metamorphosis. He had several bruises swelling his features, and Uubiron added one more.

"He is no less a victim than us," I reminded my comrade.

Uubiron bit into his bottom lip as he debated this truth internally.

"I am well *aware*," were his only words on the subject.

"Leave today as yesterday," I insisted. "We must look *forward*, to the future."

"*You* are the one still living forwards and back!" he cried.

I stared incredulously at my comrade – his words an impossible coincidence. I had not revealed my experiences in my dreams to any and it was the presence of this entity within myself I yearned to answer.

"Then I shall go *home*, comrade – if nothing else, to truly find myself," I assured my friend.

"Good," he responded. "You can take a stealth vessel. It will get you through the Break and you can pass the checkpoint with no trouble."

He turned to my human companion.

"And – Cleopatra, yes? Cleopatra, we can take you wherever you wish to go," he offered her.

Cleopatra's hair was in rough condition. She closed her eyes, dark bags underneath. Her nose was running and her clothes were stained. Any other personality would want nothing more than to return to the comforts of home for a hot shower. This woman had lost much, however, and as she gazed out the thick glass at a world beyond her Earth, she was certain her vengeance was incomplete.

"I'm not leaving you," she said, speaking directly to me. "You owe me a life!" she cried.

"A human would be easily discovered on Nemi," Uubiron argued.

"Don't give me that shit!" she cried at my comrade, who shrank away from the woman in her fury. "These people stick out like sore thumbs!" she continued, motioning to the Militia Majora soldiers that stood among the human prisoners. "*You* found a way to look like us!" she added, pointing at Uubiron.

"Cleopatra, I may never return to Earth," I informed her. "You would leave behind all you know to follow me?"

"What I *had* is gone," she stated with finality. "I want to be sure you keep your word – that you will stop this!" Tears had fallen from her eyes. She wiped them away and approached another of the human prisoners and assisted them in rising to their feet.

"Let's go," she ordered. I looked to Uubiron, who could only shrug.

I followed her to the hangar – even under such duress, she remembered the path.

We two did not speak as I prepared the ship. The silence irritated me and I began humming a tune from my youth.

"Do you have to do that?" Cleopatra finally interjected.

"Sorry," I apologized. "We're ready to go," I informed her. We strapped ourselves to our seats. I controlled the TR-3BETA as we exited the mothership and initiated the autopilot to take us to the Break.

After several more minutes of silence, I turned to my companion to attempt conversation. I found Cleopatra with her head hanging loosely. She had fallen asleep, the day's events and excitement finally causing her to collapse.

I gazed out the view port into the vastness of this realm. One could not escape the infinite points of light, each hosting its own worlds. I had feared this unknown, seeing only enemies in every direction – *any* unknown perceived as a threat to Nemitian existence.

This was not the realm I had imagined as a youngling – humanity, no *demon* that lurked in the dark.

Hours passed as I questioned whether I could identify myself as a force for good or a force for evil. I had been conditioned to believe I was working in the best interests of all – an agent of the grey.

As a light manifested itself before our vessel, a blanket of white spreading over the void, I wondered if the humans would now tell their children stories to fuel nightmares of new demons beyond the Break –

the denizens of an alien realm called Nemesis. The light consumed the ship as Cleopatra and I traversed the doorway between our two worlds.

I was toying with the damaged sprite in my pocket. I retrieved it just as we crossed the border between our two realms. I watched as it disintegrated in a bright blue flash and re-materialized just as violently as we entered Nemitian space.

My senses distracted, I did not hear the voices chittering in the light.

She has come, they whispered.

She has come.

The Bedtime Story

"Kriss! *Kriss*! Settle down! It's time for bed! Don't you want to hear a bedtime story?"

"Yes, yes, mother! Please, I'm sorry!"

"I will, but since you were bad it's going to be a *scary* story!"

"That's OK, mother!"

* * *

Contact had been lost several weeks ago.

Upon entering the area to study irregular radiation spikes, reports began to cut out due to electromagnetic interference outside the Break, the portal through which we discovered a new realm.

There was only static. Static in the morning, static in the evening, and static in your sleep. Missing, in a strange unknown space, after shipping out to identify what could be any number of dangerous phenomena – they were here to discover what lies beyond the Break and the fate of those who passed through before, never to return.

"Kaancha...," a voice called out.

Usually the demeanor of the source never betrayed his intentions. Today, from the look in her eyes, there was no hope of hearing good news.

First one up, last one down, neatly groomed, and always impeccably dressed – this was a career soldier of many years. She gritted her teeth as she saluted, always wanting to appear the professional. This was to counterbalance a few wild streaks, but there was no better friend than the ship's pilot. There was no one the Captain would rather have watch his back.

"[Captain, there is something...different, about this place,]" she remarked. "[The Break has had an effect on the surrounding space. There is spatial interference making our usual propulsion and mapping systems a bit buggy. I can see how they might have gotten lost, and once we pass the belt ahead we will lose communication.]"

A long silence followed while a plan was developed within the Captain's mind. Continuing through the interference zone would bar contact with anyone on the outside of this system, including a buoy placed near the Break. There were surely things hiding in the dark, and the Captain had no intention of risking the entire crew to suffer whatever fate befell the last.

"[The prototype,]" the Captain ordered.

"[Supe?]" the Pilot inquired, deferring to the Captain's higher rank.

"[Have us stop just on the edge of the belt and prepare the new shuttle for launch. It uses the same beacon as the one on our missing ship, and the new engine design will more easily navigate the interference. Keep the *Dark Eye* among the asteroids. As long as you remain there, we can keep radio contact, and you can link us to the buoy at the Break. If you *must* leave the belt, you will lose us.]"

The Captain had a plan.

"[I'm not coming with you, Supe?]" asked the Pilot, determined to join the mission.

"[I need you here in case the crew needs to retreat to safety in a hurry,]" said the Captain.

"[With all due respect, Supe, I think I speak for the whole crew when I say we wouldn't be leaving without you,]" came the reply. She stared back at the Captain with determined eyes.

She would not be swayed.

It was true, this crew was too stubborn and rag-tag for normal Militia operations – nor was this your usual ship. The Pilot received a quick nod from the Captain, and with the first sign of a smile since the voyage began she gave a quick salute. She disappeared down the hall towards the launch bay.

"[Just like a pilot,]" the Captain chuckled. The Pilot's last mission as a member of the Mirrit Militia was considered a tragic failure, and she had been itching for a chance to fly again.

It was cruel putting her aside to waste away on a mining platform. An aerial acrobat among the fleet, she now worked operating mining craft. It was not her choice to go on that mission – it never is a soldier's choice – yet she was blamed for the loss of her crew.

Her skin had begun to blacken in spots from radiation burns. The Militia did not take the proper safety measures and a single stray burst overloaded the crew's vessel, resulting in an explosion in the engine room.

Rather than take responsibility for sending them on a suicide mission, the State publicly shamed the only survivor. She was removed from service and the Militia no longer felt obligated to procure help for the physical and psychological effects he suffered. The ordeal nearly stripped her of her sanity.

It was the luck of this crew that their mission was nonexistent, officially. Easily gathered is a group of those needing a second chance, and the Militia will always seek out lost souls.

"[We are all born Militia men and women.]" The Captain muttered to himself, as he had been conditioned to as a child.

Gazing through the porthole, he could see the belt of asteroids they would need to traverse. Passing through this turbulent zone of the outer heliosphere, created by the sheering forces of solar winds produced by this monosol system's particle emissions pushing out fighting the weight of the cosmos, was a challenge for even the mightiest vessel – the mightiest vessel this was not.

The ship took some slight damage dodging a few small stellar objects due to the interference. This was a war of wild mountains demolishing each other, the victor charging on to its next confrontation. Kept at a distance, it was like a dangerously choreographed dance.

One must always take risks when breaking through the barriers between the unknown and the truth.

After suiting up for the expedition in the standard O-suit, the Captain looked at himself in the mirror. On the bed was his standard uniform. It was Khano Militia issue and covered in medals. Medals for acts of honor – acts of leadership – that felt more like evidence of crimes against an enemy that was just as sure in their beliefs as they were in some great war past. Each of us recruited by the Militias that served their communities.

He fought to protect *his* community, that of the Kynthra province.

"[What if there was life here,]" he asked himself aloud. "[What would they think of you then?]" He was proud of what he did for his country when he thought he was protecting his way of life. Now he sees the real victories were on paper, victories of peace and diplomacy.

Victories he had no part in.

He removed each of the medals, remembering those lives that had been lost rather than those that he had taken.

War accomplishes nothing but death, yet the Militia Majora stands to oversee the State of the Three, the treaty that has become world government. He thumbed at his Captain's pin, a gold crescent with three points bored through it, but ultimately decided to wear it. He grabbed the helmet to his rebreather and prepared himself for the journey ahead.

He was interrupted by a communique from the hangar deck.

"[Captain, that's my engine and it is not leaving without me on it! And if you think I'm going to let that joystick jockey fly *my* shuttle–,]" a gruff voice cried.

"[The *Dark Eye's* shuttle, Adunaan – and he is our best navigator,]" the Captain retorted.

"[She flies *drones* – in a mining cluster! This is the first Nemesis-class engine – the greatest achievement in trans-system propulsion since Torr's first field accelerator! It is my life's work and I will not have it flown by that *moonie*!]" The Engineer spat as he spoke the last word, a heavy insult from a man born in the upper caste.

"[I will not hear any more of this attack on my friend! She was forced to leave the service after years of flying in the most turbulent of the Aether! You will be *allowed* to come along and supervise because we will need someone to watch the engine and fix it if we run into any trouble. But you will accept my decisions on my crew and my decisions in the field on this assignment as *law*!]" The Captain's tone ensured that this conversation was now over.

The Kaancha of the Moloke expeditionary vessel *Dark Eye* observed the away team gathered near the launch bay doors. The Engineer, the Pilot, and a soldier recruited as an extra gun in case of any trouble – simply designated by her rank, Alpha – were standing at attention in their O-Suits. They all lifted their helmets – the Captain's boasting a crimson streak over the field of blue – and headed for the shuttle.

The engine in this shuttle used a combination of powerful electro-magnets and heavy coils of a gold composite for conductivity. It was dubbed the Controlled Filed Generator, or CFG.

Once the magnets were active, they push away from each other and tightened the coils. The pressure against the ship provided unfettered motion in a vacuum and an alternating current ensured swift acceleration. While fuel was not required for sheer propulsion, it was equipped with maneuvering boosters around the ship on a broken ring to alter the heading of the ship. The pit boys finished gassing it up for atmospheric flight as the Captain approached.

"[All rigged and rugged, Captain,]" one sounded off loud enough to be heard through the mask. It was the traditional ship status report. While giving a salute, oil began dripping down his forehead.

They received the nod and the pit boys were off. The Captain stood in front of his remaining crew, activating and broadcasting from the built-in headset.

"[I suppose this is where I give a grand statement about our duty – some words of wisdom that might save you out there,]" he stated, motioning towards the crackling electric blue barrier maintain the atmosphere within the hangar.

They all sat in silence while the Captain inspected the outside of the ship. He then took special care gathering ordinance and distributing it among his comrades. Three standard-issue pistols and a blaster for their escort.

As he reached out to Alpha, he noticed her hands were full.

"[If it's all the same to you, I'd prefer to keep mine,]" she said, shouldering a heavy rifle.

While focused radiation technology was the standard for weapons, this one was an antique. It was a personalized rail gun designed by Khano Industries to maximize physical damage rather than to incapacitate one's enemies. A green cloth bearing the insignia of the State was hanging from the barrel.

For a moment the Captain paused. This was a soldier brought on by their superiors in the Militia Majora, an autonomous force whose purpose was to prevent any one property from becoming too powerful. Their overreach dwarfed that of the State. She was one of only a few that had not been hand-picked by the Captain.

He began to think the State knew more than they were letting on.

Suddenly, he erupted into a boisterous laugh that reverberated across the metal floors and walls of the launch bay.

"[You can take whatever you want, there's no life out here!]" he cried, causing the crew to loosen their posture and laugh a little. They stood at attention and expected some words from their Captain.

He continued on in this silence before finally grumbling a quick, "[Let's go,]" and took his seat at the command station.

Each took a position at the control paneling around the cabin and activated their O-suits – the prototype had not been outfitted with environmental technologies. The masks tightened around their heads, the large black sensors that relayed visual information forming a seal around their eyes.

The Captain took his seat and activated depressurization. The passengers strapped in: shoulder to hip, shoulder to hip, and once around each ankle.

The Pilot's legs were free to control the pedals used to throw the vehicle's pitch. Left hand on the throttle, right hand on the joystick, her fingers ran across each handle and gripped them tightly. It felt as though she was extending the appendages of her arms.

"[Initiate the CFG,]" called The Captain.

As the first order was carried out, a low hum began to sound in the cabin. It resonated from the bottom of the ship and as the tail of the ship rose, they looked out at the space beyond the open hangar doors.

Before the *Dark Eye* were several rocks forming the asteroid belt it must traverse to gain entry into this system. They were silhouetted against the light of a single distant star. While it was known there must be planetary bodies beyond, the wall obscured any such bodies between them and this light. Beyond was a twinkling blanket of stars.

So many stars, the Captain thought as he stared in awe along with his crew.

"[So many stars...]" muttered Alpha aloud.

Those overwhelmed had only known two stars, lighting the sky by day. Reflected by the three moons, this was also the light that illuminated the night. Until the Break appeared in the abyss.

Whatever the spatial anomaly was, it continued to grow. It appeared as a bright tear in space, first discovered by telescope and driving the industry of their world. A satellite was constructed and sent beyond the threshold. The connection was lost for several hours, a mystery that remains unsolved.

When a signal was returned, sensors first identified radiation confirming a single star and possibly star system was obscured by the Break. When the first images of this new realm reached the Moloke Institute, the observers reacted similarly to the crew.

These stars had not appeared suddenly but were obscured by an aether force believed to restricting the stellar winds of the twin Suns. Curiosity grew as they gazed into a new and vivid cosmos which had fueled industry with dreams of reaching it. They could only imagine what lay outside the small lens of the Break.

"[Keep a cool head, soldier,]" chortled the Captain. "[Think of the beauty we can find at home – may find entire worlds here more beautiful. Save your awe for the end of our journey.]"

The Pilot had been slowly maneuvering the prototype – the *X-1* – out of the shuttle bay. The Engineer had been watching her closely while tinkering with some figures on his LightPad. They were in position one kilometer from the *Dark Eye*.

The Captain sat up sharply and considered the burdens of this crew. Not the first to see beyond everything known to exist, but it is unknown whether those who came before will ever speak of what they found.

It is unknown whether *this* crew will meet the same fate.

He assumed the State would symbolically memorialize them all with Honors posthumously, erect a monolith to commemorate the tragic loss, and never again venture beyond the comforts of the known.

The alternative was home never learns of what happened here, continuing a history of government secrecy and disinformation. That was a weight he did not wish to add to his coffin. His crew expected reassuring words. They too felt the pressure of their success – or failure

"[The time for speech has passed,]" he stated.

"[Hit it.]"

They were pushed back gently in their seats. While the CFG engine provided the artificial gravity within the cabin as well as the power for the main propulsion systems, it did not completely negate the G-forces of accelerating to speeds previously unreachable. They quickly cleared the few hundred kilometers to the rock barrier. The Pilot pulled back and the *X-1* began to climb along the height of the wall.

"[Sensors are reading a few larger bodies but mostly formation debris. Wow…it's amazing how much material is here,]" said the Engineer. "[The Regent would be very interested…,]" he began.

The Captain cut him off, asking, "[How long?]"

"[We should clear the thick of it in a few hours, Supe,]" the Engineer replied.

"[Unacceptable. I guess our navigator is going to have to put your ship to the test,]" quipped the Superior, giving the Pilot a nod. Before the Engineer could retort, she dove into the asteroid field.

"[Whoo!]" she cried gleefully. The Engineer held tightly, Alpha seemed unaffected, and the Captain let a grin creep up one side of his face. Someone had to push the limits of this design while its usefulness was still under scrutiny, and there was always time for a little fun at the expense of the Militia.

They watched through the thick glass as the craft whipped between the asteroids. Relatively speaking, the motion of the rocks and ice were slow, though any collision would surely obliterate the ship.

As the Pilot navigated beautifully through the chaos, The Captain thought to ask The Engineer to find appropriate music.

The belt was three times as deep in this direction as it was tall. While dangerous, this path was the shortest distance between us and the inner system, parallel to the plane of the star and any planets within. A moon-sized spherical body was detected but it was unlikely to support any form of life.

They exited the field of goliath masses and found themselves within the interference zone. The Engineer calibrated the sensors to detect the missing ship or any ejected escape pods.

They found nothing.

"[I've detected a gas giant within the inner edge of the field,]" piqued the scientist. Hours had passed, and Alpha gripped her rifle tightly, preparing herself, bored of the journey. Her mind was filled with thoughts of the possibility of hostiles.

"[Atmosphere of E-1, E-2, and C-6114,]" The Engineer continued. "[It seems to have water, icy C-7113. Solid core, but I can't say much for a surface. *Supe*,]" he added as an afterthought.

"[Search for any signal outputs,]" came the Captain's voice. The only inconvenience had been the interference of the EM field beyond the wall, but there should be no obstacles for the sensors here.

"[Check the moons, any rings that might have interfered with their thrusters, clogging them up,]" he ordered. Several minutes passed and the planet came into view.

The planet was stunning, and they were all speechless as this first vision of a new world grew before them. It appeared as a solid blue orb, streaked with white clouds like a Kynthra pearl in pristine condition. As younglings, all knew how to shoot them from between their finger and thumb to knockout an opponent. It was a common past time.

"[Nothing, Captain,]" came the Engineer's worried report. "[The *X-1*'s sensors have not been calibrated for range, and I can only see this side. There may be a way to extend the…,]" he began murmuring to himself.

"[Do what you must,]" sighed the Captain. "[For now, circle the planet, Pilot, and keep the receivers on.]"

While there were no outstanding rooms, there were two benches that could be used as beds. The Captain looked around at his assembled crew as he sat down. They were strapped in except for Alpha, who was at full attention.

"[Alpha, loosen up – this isn't official Militia business. Not unless we're dead,]" he said, catching her off-guard. To this soldier standing at the weapons console, he said, sternly, "[And *don't* say hello without me.]"

They continued after finding no success at the first planet. While this ship was fast, it was not equipped for great distances. It would take days to reach the inner solar system. They came around the horizon of their first query reveling the background of more lights than they ever believed could be in all existence.

There were few that were brighter than the others and had a hue to their glow. These were more planets, three visible between the ship and the second asteroid belt within the system. This structure was large enough to darken the star and be visible from The Break, though it would be much less of a challenge than their first obstacle.

The planets were most likely gas giants similar to the one their home world revolved around. It may be a moon, but it had no shortage of wonders.

They thought of the great cities they had built over millennia on a world of vast colorful forests, great mountains, and deep oceans. A vast variety of life roamed the earth. They wondered what they would find familiar and what they would find alien here.

Much on this side of the Break was already so different.

Most join the Militia to see as much of home as possible, climbing the ranks to gain access to areas of the planet restricted from civilians. Plains of volcanic eruptions, a source of thermal energy, powered a continent. Outposts were constructed on the other two moons to test our ability to survive. It all belonged to the Militia.

Those that fought in wars past knew their place would be in battle. However, to be allowed onto the stations between the moons, exploring the other worlds surrounding their planet – their garden, and their guardian – was the dream of many younglings.

The first scan was completed and they quickened towards the next celestial body. They were lucky to find the planets in this system to be in near alignment.

"[It's like they were expecting us,]" joked the Pilot. "[Judging by their orbits, the journey should have been as smooth for the science crew.]"

"[We can use these planets' gravity in order to boost ourselves even more quickly to the inner solar system where we are more likely to find *hospitable* conditions,]" added The Engineer. "[That is where they would have been searching – I'm positive.]"

"[Shouldn't we run that by the Captain?]" the Pilot attempted to interject. Her crewmates looked at each other, then back at her, smiling.

"[What the Captain is blind to –,]" began the Engineer.

"[– won't harm him!]" finished Alpha, eager to be of use. As they dived towards the planet, skirting the atmosphere before blasting back into open space, the maneuver caused turbulence within the ship. The Captain shook on his bench, but did not otherwise stir.

The Pilot remained suspended in the cage. Alpha played a game default to the operating system of the LightPads. After several hours, the Pilot required relief. This relieved the Engineer, who had been buried in the data generated by the sensor panel.

The Pilot slipped to the back, careful not to wake The Captain. She could not help making a face to test if he was truly asleep.

Old habits die hard between friends.

Their next stop was no less astonishing to the waking crew. Alpha retrieved the Pilot to man the intricate maneuvers around the planet, a similar hue to the first. Its axis was perpendicular to the plane of the previous bodies within this system and it had thin rings that required an expert touch to navigate. While searching for any signs of their lost comrades, the Engineer was needed to read the sensor panel. They heard a shuffling as the Captain too awoke to find the new planet outside the viewport.

The Captain grumbled a request, expectant of an update. The Engineer, startled, looked down at the readout and shook his head.

"[Then we continue the search,]" the Captain said, leaning back into his chair. "[Have you made any progress with the sensors? What's our range?]"

"[We will be able to ping the whole planet and the moons around it within an hour,]" the Engineer informed him. He took this opportunity to relay the results of their plan to slingshot around this planet and the next. The Captain agreed – the sooner that they confirmed the previous expedition had survived, the sooner more of their people could come through The Break.

It was possible they would find a fertile world to populate. There were many reasons one would wish to leave life in the Nemesis system.

There were many mistakes one could be running from. The *X-1* abruptly came to a stop, this time startling the Captain. He looked up and saw what had caused the Pilot to halt the craft.

The next planet, another gas giant, had two thick magnificent rings surrounding it. So great were the rings that the planet cast a shadow that spanned only half their radius. The atmosphere was turbulent and appeared in many bands of different colors representing the makeup of gasses swirling about the planet.

"[What have we got?]" inquired the Captain.

"[Uh...C-12, E-2 in the atmosphere –,]" the Engineer started.

"[Not what I want to know,]" stated the Captain through gritted teeth. It became apparent he was not a morning person.

"[Supe, all within range and reading 53 moons. It will take a moment to search them all, but some of these are similar to home. I can't tell if there is any life of the surface, but...it's *amazing*!]" the Engineer cried, never taking his eyes away from the information on his panel.

After completing a sweep of the moons, they began procedures for the slingshot maneuver. The rings prevented them from approaching too closely, but provided a stunning image as they passed.

Much of the data they retrieved had been visual documentation for the scientists back home to scrutinize more thoroughly.

Alpha, a soldier but no simple mind, thought of the art her people would create. These imaginations of the many worlds beyond the Break would inspire the most beautiful paintings, the most wonderful stories, and even the most exciting of music.

The largest planet yet, more like that of Nemesis itself, began to grow before them until it filled their view. A giant storm of swirling red gasses was raging in its atmosphere. More worlds reminding them of home were made of its moons – however, the sensors still received no signal. They read no debris among the surrounding ice and rock of the planet's thin rings.

The Pilot again swung the ship around for their last leap into the unknown. If a gas giant was on the other side of the approaching asteroid belt, none were apparent. Any other planets beyond would be completely foreign to their eyes and senses. They angled their descent to this large planet so as to be flung below the next obstacle – the inner asteroid belt.

They had been flying for three days. No sign of the missing vessel and contact with the *Dark Eye* had been lost. It was assumed the bodies pirouetting behind them provided too much interference – the staging vessel was no longer receiving any scan data.

As they came about from under the shadows of these last few obstacles, The Pilot pulled up such that all the space between the asteroids and this star was visible. Four worlds shined before them, none of them gas giants. No one had ever seen anything like it.

"[Sensors?]" asked the Captain.

The Engineer had improved their sensor range in the time it took them to reach the inner solar system. He began to read out what they were picking up. Twice the size of the moon they lived on, rocky worlds alive in activity but for the cold red planet nearest them. Each of the four planets were again aligned, begging to be explored. The previous expedition could not have asked for better conditions.

"[Any sign of them, soldier?]" the Captain instructed as he motioned for Alpha to check the comms panel.

"[This is Team Rescue-1 on all frequencies – is anybody reading?]" Alpha called into the silence of space. She then switched to receiving and the most peculiar sounds began to flood the cabin.

"*Rinso Presents: Call the Police!*" they heard a voice speaking in an unidentifiable language. "*Attention Homicide Sergeant, Crime Squad Detail. Murder suspect in your zone, close in according to instruction.*" What they could only assume was a form of music followed with the squeal of alien instruments, and a new voice came to their ears that was no less impossible to understand.

"*Between you and the evil outside the law, between you and the housebreaker, the kidnapper, the murderer, stands the Policeman of your-*"

Alpha cut the audio. There were other voices coming in on different channels and overlapping each other. It was unclear to any on board whether they heard the same language or not. This was unexpected and they didn't bring along a translator, though it was more likely they needed an anthropologist.

"[I'm sure the previous crew heard them too,]" chimed in the Pilot. "[Do you think they would have investigated? Why wouldn't they have reported this first?]"

"[Maybe they couldn't,]" the Captain argued, "[We don't know at what point they became immobilized. It is possible they tried going through the asteroid field and became damaged. Any read or response? Where is this signal coming from?]" he inquired of the comm.

"[Third rock from the sol, Supe,]" came the soldiers reply.

"[But no sign of our people,]" the Captain stated plainly. No one spoke for several minutes.

"[Then I guess it's time to say hello. Forward on,]" ordered the Captain. These voices had not responded to their original hail. It was possible they did not receive it, but their level of technology remained an unknown foregoing a tighter approach.

No ships had been detected. They could assume these were not travelers of space, though a few pieces of debris appeared to be floating just above the atmosphere. The limited amount of this debris calmed their nerves.

These people – if these were, indeed, *people* – had primitive technology. They conjectured the previous ship did not encounter any trouble from the planet but possibly landed to seek help in desperation.

They began their search and found something most unexpected. Any number of locations across this planet could potentially be the source of the random bursts of radiation they had been reading through their lens of The Break, inciting their curiosity in the first place. The first crew had found what they were looking for.

"[Captain, we've got it,]" the Engineer said as he looked up from the sensor panel. A piece of the hull was found in a field about 50 km south of a nearby city.

They studied the area closely. The air was thick with emissions from crudely fueled vehicles as well as residual radiation from whatever war the aliens must have been fighting.

These new visitors were reminded of their own history. Many wars had been fought over possession of land and resources. This world seemed so full – so giving – one would wonder how these beings could fight over it, risk destroying it, and risk destroying themselves.

"[Survivors?]" the Captain questioned half-heartedly.

He received the answer he expected.

"[We'll need to find the brick. What part of the hull?]" he continued.

"[It's the cabin, Supe,]" the Engineer responded. "[We'll have what we need.]"

"[And we can provide them The Rite...,]" muttered the Pilot. It had been a long-standing tradition that the deceased were sent through the Break.

The CFG was deactivated, maneuvering thrusters capable of operating the ship within an atmosphere. They were careful to hide behind any clouds to remain unseen by those that would be their hosts.

The *X-1* was both an engine and defense prototype, and its next function would be a great asset, should it work.

"[Activate cloak,]" ordered the Captain.

The outside of the *X-1* shimmered a bright gold and faded until only a slightly distorted version of what lay beyond it remained. The moon was visible through the cloak, holding itself in the night sky in a waning phase – twenty percent luminosity. Confident they would not be discovered, they landed and moved onward to the crash site only to find lights in the distance approaching more quickly.

These beings were arriving in vehicles to inspect the site. The crew of the *Dark* Eye would be unable to reach it in time, though only they knew what to look for. They resigned to a vantage point a short distance away and watched as the creatures began to pick apart the fallen ship.

Once the beings had left, the Captain quickly rushed down the hill to the crash site. The LightPads showed that the brick was still intact and broadcasting, containing any information that had been gathered on the previous expedition's journey.

They could not find brick among the wreckage – even more dire, they found no bodies. They had witnessed transport driving away from the site and had assumed they were taking materials for study. It was becoming clear they also took any remains of their people.

While they had held out hope of finding survivors, with no knowledge of their potential captors they were unsure if it was possible any of the previous crew remained. They formulated a plan to retrieve the brick and their people from a base the vehicles were headed towards, following their tracks through the dirt.

The journey would take too long on foot, prompting the Captain to retrieve the AirMo, a single passenger vehicle that sat several inches off any surface, while the rest of the crew retreated to the *X*-1. They established a temporary base of operations.

As he twisted the throttle, the smaller CFG module whirred to life. The scooter glided silently and effortlessly across the desert, following the tracks made by the transports until reaching what looked like an airfield.

This world was capable of flight, though they were utilizing pre-CFG technologies. Space travel was likely out of their reach

While established hangars and warehouses were positioned along the landing strip, there were tents set up in a rushed fashion located away from the main structures. It was this direction the Captain was headed.

Evolution allowed him to change the color of his skin, as well as the texture, when necessary. It is believed to be a remnant of a defense mechanism used to camouflage an early ancestor, though most Nemitians had lost control over this phenomenon. It now reacted subconsciously to emotional stresses, blushing green when embarrassed or blue when in a rage.

As the evolutionary chain continued, and the masses lost their control, other mutations empowered these genes to allow near complete invisibility in those that could express them. His blood was translucent as well, aiding the effect.

It was through study of this expression that led the development of the *X*-1's cloaking technology. The Captain was of a lucky few with this ability and activated his natural cloak to explore the make-shift base.

A closer inspection of these peoples' technology and infrastructure resembled the progress made at home two hundred years ago.

While using crude fuels to power their vehicles, there were devices here to suggest they were familiar with harnessing atomic power at some level. From what data they had collected, he was not surprised to find they used it for weapons of mass destruction rather than utilizing its potential to provide this world limitless power. Coincidentally, the standard issue rad pistol on his hip was of similar design

They were abusing a much larger planet than the moon-world, Nemi. Saltwater covered the surface ripe to be filtered and vast terrain full of resources that could save his world stretched far across this continent. Judging by the images of cities collected from orbit, these beings were severely over-populating their world beyond its ability to support them.

Some were wearing rad-suits, reading levels emanating from the debris they had recovered. A string of clicks rang out as they walked closer towards the control panel of the craft. While the prototype did not require nuclear power, this was a standard Moloke Investigations vessel.

Just below the control panel was the engine. The floor was an impenetrable composite protecting the cabin from the high intensity radiation. The floor had been shredded, exposing the engine housing, but it was no longer active.

It was here the Captain could also see the brick, a black box hanging loosely by several severed cables. The investigators took no notice of it as they loaded it onto a cart.

They removed the atomic material and placed it in a large barrel, surely of a similar composite as the floor of the crashed ship. While he could not understand their language, it was apparent they were an intelligent species, their society likely having evolved over the past twenty thousand years, approximately – just shortly behind his own. This system provided them boundless resources if they someday learned to navigate it.

The Captain did not wish to linger here as it caused a stress on his body to hold this form. He could not bear the stench of these creatures. As this was the third planet, he began calling them *Loarkeen*, from the word *loar*, meaning, "[three,]" and the word *kincheer*, meaning, "[smell.]" They had only been viewed from a distance or in full body radiation suits, making this their most defining feature.

He wondered if an anthropologist would agree.

As the Captain moved along the tents, his eyes were drawn to several layers of what he recognized as decontamination protocol. Something was drawing him in this direction. He waited until there were only a couple of these Loarkeen about.

These two were in the distinct garb of military operations: camouflage fatigues, helmets, radiation masks, and guns. It was a look he was all too familiar with. He held his breath as he walked past them, nervous for the first time after entering the camp.

A familiar scent assaulted his nostrils here, emanating from just beyond the entrance.

Burning flesh, unmistakable on any planet. While powerful, it was not enough to overcome the terrible stench of death and rot. He did his best not to start running through the walls of this tent.

These aliens were wearing no suits – no masks. They walked with charts and carried many instruments. If his anxiety had not taken hold, he would have studied them more closely with his LightPad and gathered information to provide the Moloke Institute.

He quickly shuffled through the weak-walled hallways as if a light breeze, pushing aside the sheets to avoid bumping into these scientists. After two partitions he found his fallen comrades.

* * *

"And that's all you will ever need to make a little younglings' nightmares!"

"Father, I'm not scared! I know, someday, I'll get to fight the demons! And I'm going to kill them all!"

"We know you will, our *sweet* boy."

"We're all born Militia men and women!"

"Yes we are!"

"Goodnight, mother – father!"

"Goodnight, *Kriss*."

"..."

"Well, don't stop on my account – finish the story!"

* * *

The scene before him was a gory mess of exposed organs, removed from the bodies and studied on a table surrounded by many of these people. A lifeless head rested on the table looking back at him. He did not recognize the face, but it made no difference.

Whatever similarities they shared ended at their appearance. While those around him had smooth skin and tufts of hair on their bodies, the corpses before him were scaled – hairless. The eyes of the face now before him were beads within muscular eyelids that could move independently of each other. They were cold and the crest mounting the skull deflated.

He began to feel a beating in his skull. This was not a proper way to treat a body. He could compromise with an ignorance of The Rite, but this was a mutilation. The crew were not animals to be dissected and studied. He could not help but judge this species as wholly hostile. He would see them pay for this affront to his people and swore that they would suffer.

His rage grew as he thought of what numbers must live on this planet. He found his hand on his weapon – the pistol he had at his side – and decided these deaths would not be enough. He turned to run, knocking over two scientists in his path. He ran across the mass of tents, across the airfield, and behind the base where he had left his AirMo. He flew back to his crew and told them what he had found.

"[All they would have wanted, all they *deserved*, was The Rite,]" slurred the Pilot. As was custom, every ship came with a bottle for such an occasion. It was probably best the pilot not have more, even though this drink was specifically brewed to be a short-term intoxicant.

A more important custom, The Rite, was at the core of their religion and mythology. They believed while the body may be destroyed, the soul can exist within the Break forever – outside of space and time. This is where the spirit of those that first came to Nemi lived with the Ancient One. Their souls were believed to pass on to the What-Comes-After, the crew thought confirmed by the lack of corpses floating on the opposite side of the portal between realms.

The Captain took the bottle and finished it. He spent his ten minutes outside of sobriety silently watching over his crew preparing to finish their night at the edge of camp. It seemed the Pilot had brought a second bottle. Before the crew could open it the Captain appeared looming over them.

"[We're not staying,]" he informed them. "[Phalae Alpha, Yekk, you're with me. Everyone else – pack up. We have to take what we found back to the State. We have war to make.]"

The Captain took Alpha to a farmhouse nearby. He had gained intelligence that part of the ship had been found earlier that day by a farmer and his son. While these militants had surely taken any material they could find at the main crash site, it was likely this discovery was not reported immediately. One might consider such an item a treasure to be hidden away.

While searching, the Captain agreed the same could be said of home. Items from this world would surely garner envy. He grabbed a heavy leather-bound book that seemed to be the only literature in the house. He thought it must be an important tome as it was on display.

He exited and saw Alpha waiting in the brush, having found a buried treasure of her own in the form of the wreckage. They quickly returned to base camp.

They disassembled the bunks they had previously been resting the week's travel off in. Nothing compares to a night on solid ground, even if on a strange planet. Quiet nights were a luxury in the Militia.

Their temporary command center had been loaded in the storage pods along the side of the *X-1*. The crew boarded, strapped in, and prepared for launch.

The Captain stood outside, beyond the wing of the vessel. The moon was now under the horizon, but it was still night. He could see all the stars in the sky, possibly all holding new – as he now believed – hostile worlds. He boarded the ship and took his seat, making a silent vow to protect his home from these demons inhabiting this realm

"[Let's go,]" he ordered, taking his seat.

"[Captain...,]" the Pilot started.

"[What is it?]" The Captain snapped. He was ready to take leave of this planet.

"[What about the brick? You said you found it?]" she inquired. "[We could really use that information on the planet – these people – if they are to be our enemy.]"

The Captain remained silent. He was always calculating – one hundred steps ahead of the average – and this moment was no different.

"[Don't worry, Kriss – we'll be *back*,]" were his final words before lift-off.

The *X-1*'s engines hummed. It oriented itself, and took off into the twinkling night sky, fading as though another star gone out. The crew silently raced back to the outer asteroid belt, no longer in awe of this cruel world they had discovered

They returned to the *Dark Eye*, no more conversational.

They thought only of this new and unknown enemy. They had known no other life than that of Nemi – the two suns and empty skies. They believed life to be a gift at any evolutionary stage, from the microbial to themselves. Some thought the power of life to survive and adapt was the equivalent of magic.

What followed this journey were tales of the worlds beyond the Break – myths of terrible demons that were to be hunted down and smite from existence.